# Texas Rattlesnake Tales

Best Always to
"Michele" a
"my "friend"
Tom Wideman —
- 2016 -

# Texas Rattlesnake Tales

Tom Wideman

Cover photograph by Gerald Ewing

State House Press
McMurry University
Abilene, Texas

## Texas Heritage Series
Number Eight

**Also in this series:**

*Christmas at the Ranch* by Elmer Kelton

*What I Learned on the Ranch* by James Bruce Frazier

*The Parramore Sketches* by Dock Dilworth Parramore

*A Small Town in Texas* by Glenn Dromgoole

*Biscuits O'Bryan: Texas Storyteller* by Monte Jones

*Buffalo Days: Stories from J. Wright Mooar* as told to James Winford Hunt

*Back Then* by Archie McDonald

*To June, the string who anchors this red balloon!*

Jane Wideman

**Library of Congress Cataloging-in-Publication Data**

Wideman, Tom.
Texas rattlesnake tales / Tom Wideman.—1st ed.
p. cm.—(Texas heritage series; no. 8)
ISBN-13: 978-1-933337-02-9 (pbk.: alk. paper)
ISBN-10: 1-933337-02-8 (pbk.: alk. paper)
1. Rattlesnake hunting—Texas—Anecdotes. 2. Wideman, Tom.
I. Title. II. Series.

SK341.S5W53 2006
799.2'579638—dc22

2006012167



State House Press
McMurry Station, Box 637
Abilene, TX 79697
(325) 793-04682
www.mcwhiney.org

Distributed by Texas A&M University Press
www.tamu.edu/upress • 1-800-826-8911



Printed in the United States of America

1-933337-02-8 paper

Book Designed by Rosenbohm Graphic Design

# Contents



Photographs are from the author's personal collection unless otherwise noted.

# Chapter 1
# Fangs of Fire

A rattlesnake's bite compares to being stabbed with two redhot ice picks.

Adding insult to injury, nature provides rattlesnakes with excellent natural camouflage, making them difficult to avoid. Rattlesnakes strike in response to movement, heat, and as a last defense. They don't shake their rattles to warn predators—it's an involuntary response to danger. While shedding its skin, a rattlesnake is particularly unpredictable and dangerous, striking randomly without provocation.

A rattlesnake's venom destroys cell tissue, resulting in open, draining wounds that don't heal without medical attention. As a Sweetwater teenager, I worked for Mrs. Sunny Cook on her ranch in nearby Maryneal, where I often found dead sheep and goats

*The snake man: Tom Wideman. (Photo by Gerald Ewing)*

with portions of their heads rotted away—victims of rattlesnake bites.

In 1973, I owned a horse named Cowboy that I boarded on some land south of Sweetwater. One morning when I went to feed him, the area above his mouth was the size of a basketball. Upon closer inspection, I found two fang marks far enough apart that my thumb wouldn't cover them. The rattlesnake that bit him had been a big one.

I loaded Cowboy in a horse trailer and took him to Dr. Buddy Alldredge, Sweetwater veterinarian. I was concerned that Cowboy's nose might slough off, but Dr. Buddy reassured me that I'd gotten the horse to him quickly enough to prevent that. After a two-week stay with Dr. Buddy, and massive doses of penicillin, Cowboy recovered. The vet bill was $400 back then; treatment is much more expensive now.

Cowboy was never the same after being snakebit. He had never paid much attention to ground noises, but after his encounter with the rattlesnake, if we happened to brush against a yucca plant filled with dried seeds that rattled a bit, he would be out from under me in a second.

One day Jack Stein and I were quail hunting on a cold, clear day, using Jack's dog, Bob, an expensive, full-blooded pointer. With Bob running ahead of us, we approached an old barn where we usually found a large covey of quail. All at once, Bob began howling and rubbing his head in the dirt. As Jack and I neared the dog, we heard a snake rattle behind us and spotted a five-foot snake.

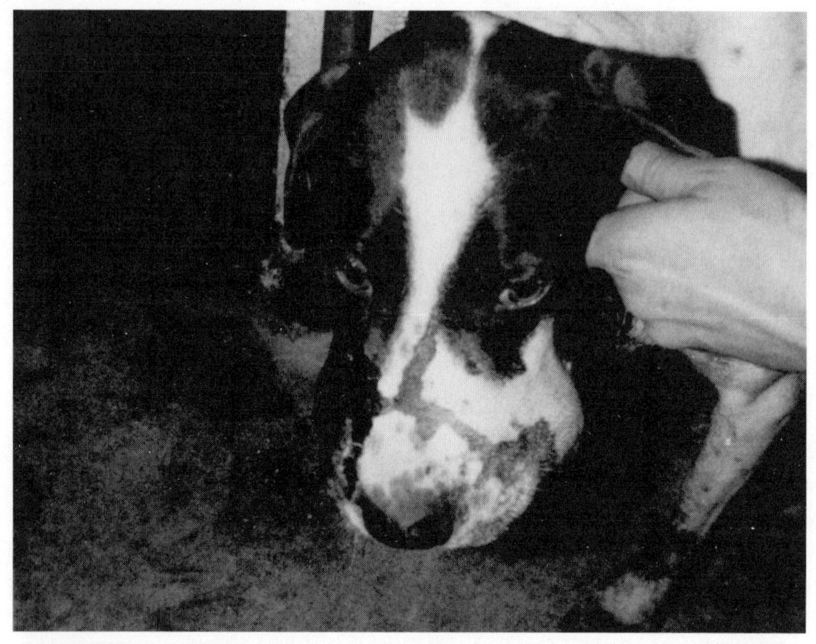

***Bob's head swelled up and he almost died after being bitten by a five-foot rattlesnake.***

I yelled at Jack not to shoot; I wanted to catch it. Then I turned and saw the dog's face with blood pumping from a vein with each heart beat. In an instant I emptied my double-barreled twelve-gauge into the rattlesnake.

I gathered Bob in my arms, and we ran back to Jack's pickup. The dog was bleeding profusely, and we were more than thirty miles from Sweetwater. I glanced at the speedometer. Jack was driving over ninety in a fifty-five mph zone. As we topped a hill, we met a Texas Highway Patrolman, who turned his patrol car around and with lights flashing, pulled us over to the roadside. Jack jumped out and ran to the patrolman's car. When he explained our predicament, and the patrolman saw the dog covered with blood, he told us to get going and offered us an escort.

Away we went to Dr. Buddy Alldredge's veterinary clinic. Dr. Buddy confirmed our worst fears when he discovered that the rattlesnake had hit a vein. Bob's head had swollen so badly that it resembled a bulldog's. Amazingly, six weeks after Bob recovered, we took him on another quail hunt, and he wasn't the least bit nervous.

Jack was worried about his other dogs getting snakebit, so he took them to a man in Merkel who trains dogs using live rattlesnakes with their fangs removed. But fangs grow back in a short time, so it's a dangerous technique. He places a shock collar on the dog along with a long leash. When the dog nears the rattlesnake, the trainer presses a control button, and the dog receives a mild shock. After being shocked several times, the dog associates the shock with the sound of the snake's rattle and won't get within ten or fifteen feet of it.

Animals bitten by rattlesnakes on the head or legs usually recover. However, if an animal is bitten on its chest or rib cage, the swelling and tissue damage associated with the bite can prove fatal.

C.H. Boney, who lives in Abilene and farms near Hawley, telephoned one afternoon and invited me to his farm to see if I could locate a rattlesnake that had bitten his dog, a small rat terrier. The snake had struck the dog between the eyes, causing its head to swell to the size of a softball. I invited my friend, Bill

Whitaker, then feature columnist for the *Abilene Reporter-News,* to come along.

After seeing the barn, I told Mr. Boney that he had created a rattlesnake "condominium" because of the old wood, broken concrete and tin littered about. Bill and I inspected the barn inside from one end to the other. I found several markings in the dirt where snakes had been, but didn't find *the* rattlesnake. I sprayed gas vapor mist several places and waited—no success.

Mr. Boney was extremely anxious to locate the rattler because his barn was unlighted, and he was afraid of getting bitten. Bill and I went outside and noticed two large slabs of broken concrete; however, my sprayer wouldn't reach far enough under them to drive out a snake. Mr. Boney had a tractor with a lift he used to move hay to his cattle. I asked him to lift the slab so we could look under. It was cold that morning, and I believed that a rattler might be dozing under the concrete.

He slid the long forks under the slab and slowly lifted. When the slab was about two feet off the ground, we spotted four rattlesnakes and a bull snake. Bill and I captured them and put them into my snake box.

I told Mr. Boney that the bull snake was harmless, but he said he didn't want any snakes around and if I didn't take it, he would kill it. He said the snake had bitten his dog without warning—no rattling sound. He was surprised when I showed him the reason why. Rats sharing the snake's living quarters had chewed off its rattles while the snake was hibernating—a common occurrence.

# Chapter 2

# Where Distance Isn't Measured in Miles

The vast size and rugged terrain of West Texas can be overwhelming to people who haven't lived there. With miles and miles of nothing but miles and miles and cloudless skies, you gaze fifty or a hundred miles to the horizon with nothing to obstruct your view. I'm partial, of course, but you haven't really seen a beautiful sunset until you've enjoyed one in West Texas. In this area, we don't measure distance in miles, but in hours.

*The Western Diamondback Rattlesnake thrives in rugged West Texas.*

Western Diamondback rattlesnakes are as much a part of West Texas as cattle, oil, and cotton. People from other parts of the country have complained that Western Diamondbacks are in danger of extinction because of rattlesnake roundup events, such as the one in Sweetwater. Well, the Sweetwater Roundup has taken place annually since 1958, and at the end of each three-day event when we total the number of rattlesnakes captured, it averages 3,500 to 5,000 rattlesnakes, weighing three or four pounds each. We would have to tame the rugged West Texas countryside to catch all the Western Diamondback rattlesnakes, and that's not going to happen anytime soon.

For at least seventeen years of the Roundup I hunted on one ranch, catching at least a hundred pounds of rattlesnakes from the same dens every year. That's just one ranch in one county. There are countless dens on this ranch that we've never worked simply because we haven't been able to find them.

I hunt for the fun, excitement, and companionship of people I'm involved with every year. It's not my intention to hunt rattlesnakes until they're candidates for the endangered species list.

People I take on rattlesnake hunts may be nervous at first, but by day's end, they're more comfortable visiting areas where poisonous snakes are found in their natural habitat. Although the rattlesnakes we capture during the Sweetwater Roundup aren't released, hopefully people who go hunting with me come to realize that you don't have to kill or keep every snake you see or capture, or that "the only good snake is a dead snake." Because of a natural dread of reptiles and the inability to differentiate between nonpoisonous and poisonous species, people seem to want to kill *all* snakes they come across.

Every day during the Roundup, twenty or thirty nonpoisonous snakes—bull snakes, coach whips, chicken snakes, hog-nosed snakes, king snakes—all harmless to humans, are turned in with the rattlesnakes. Nonpoisonous snakes are segregated and used for "photo opportunities"—hunters who want to have their pictures taken holding a snake to show friends back home. When the Roundup is over, the nonpoisonous snakes are released into the wild, unharmed and, maybe, a bit wiser for their experience.

Not much publicity is given to the hundreds of nonpoisonous snakes released back into the wild each year after the Roundup, and many rattlesnake hunters aren't aware that nonpoisonous snakes they bring to the Coliseum aren't counted in the Roundup's total poundage. Each year Jaycees in charge of the weigh-in pit sort out these harmless snakes and place them in my wooden snake boxes at the Coliseum. By the close of the first day, the pit manager may have set aside as many as twenty-five nonpoisonous snakes, which I release wherever I happen to hunt the next day. The second day, another box is filled, and I release those snakes. I've done this for more than fifteen years, and the nonpoisonous snakes I've released on rugged ranch land probably number in the thousands.

Rattlesnakes eat rats, mice and other small animals, and nonpoisonous snakes that pose no danger to man or other animals do the same. Most people's first impulse is to kill all snakes and sort them out later, but I don't kill rattlesnakes. Each year on Saturday morning—the second day of the Roundup—I turn in the good-size rattlesnakes I've caught and release the small ones—two feet or less—at a den on property where the landowner has given me permission to do so.

***Gary Wideman plays with a non-poisonous coach whip snake.***

In 1990, several out-of-state environmentalists attended the Sweetwater Roundup in opposition, but there's no way the Roundup can reduce West Texas's rattlesnake population to an endangered level. I believe our area would have been overrun with rattlesnakes if not for the Roundup.

In the 1950s, ranchers in Nolan and Fisher Counties contacted the Sweetwater Board of City Development and asked for their help in dealing with the rattlesnake population. The West Texas area is too vast and untamed to suppose that we could *control* the snake population and eradicate them.

Most roundup opponents live in cities and states where rattlesnakes aren't native. I live in Abilene, a city with a population in excess of 110,000, and at least ten people a year require medical attention for the effects of a poisonous snake bite.

My friend, Clint Arnold, was bitten by a rattlesnake in 1994 while taking a water meter reading on his ranch northeast of Abilene. He saw the snake and was trying to hit it with a brick when the snake struck him on the finger. Because the wound bled profusely, he and his wife, Esther, thought that the venom was draining out. They didn't try to extract the venom or use a tourniquet to limit its spread. Within minutes, however, the bite began to cause Clint great pain, and Esther drove him to an Abilene hospital.

Doctors in the emergency room administered a hypodermic shot of anti-venom serum. Unfortunately, Clint had an adverse reaction to the anti-venom, and his blood pressure dropped to almost zero. After a few uncertain moments, his condition stabilized, but he spent a week in the hospital and several more weeks sidelined at home because of the snakebite.

Coexisting with a large rattlesnake population will always be part of our West Texas lifestyle. They were here long before we were and don't take kindly to our urbanization of their habitats.

# Chapter 3
# Pick Up What? Using What?

If you're going to adopt rattlesnake hunting as a hobby, the two most important hunting accessories you'll need are a cool head and a deep respect for the critters.

Rattlesnakes are extremely unpredictable in their natural habitat, and there are lots of them in West Texas dens. I've never been afraid of rattlesnakes, but I have a very healthy respect for them.

Be on guard when you're snake hunting or in an area that snakes may habitat. Rattlesnakes have excellent camouflage. When you're out in the field, never put your hand near the ground without first looking around. I wear unlined leather gloves and a loose-fitting, long-sleeved shirt while snake hunting. Go slowly; pay attention to your surroundings; look and listen. Sound a bit like instructions at a railway crossing? Well, if you're bitten by a rattlesnake, you'll feel as if you've been hit by a train.

If you see a rattlesnake or hear one rattle a warning, freeze in your tracks. I know that's hard to do when your natural inclination is to run, but you should remain absolutely motionless until you determine the rattlesnake's location. Chances are that if you hear a snake rattle and don't move, the snake will retreat first.

Once you determine its location, back away slowly at least three or four feet, watching carefully where you're stepping. Rattlesnakes strike only one-half or two-thirds of their length—usually one-half. You're more likely to be bitten if you panic and run.

Another important reason for standing motionless when you hear a snake's rattle: there may be more than one rattlesnake and if you back off in a hurry without looking, you may step on the snake's traveling buddy. Use common sense if you're going to hunt rattlesnakes, and even if you never actually hunt them, remember to stand still if you come upon one or two of them in your path. My wife, June, made a deal with rattlesnakes years ago—she won't hunt them if they won't hunt her!

Don't hunt rattlesnakes without reliable snake tongs at least thirty-six to forty-eight inches long and a strong wooden box or metal container. The best snake tongs are factory-made from die-

***Tom Wideman demonstrates his snake tongs.***

*The author has used this snake box for more than twenty years.*

cast aluminum and can be purchased in most hardware stores. The snake tongs I use have a set of finger-like holders activated by a trigger handle and are always available at the Sweetwater Roundup. I know several people who make their own tongs, but they have the appropriate manufacturing equipment.

Some snake hunters use a snake "noose," which I despise. The noose is made using a piece of conduit with a long wire run through it, looped on the end. Using this apparatus is dangerous for the hunter and usually kills the rattlesnake when the noose is pulled tight. The rattlesnake goes crazy and breaks its neck.

A rattlesnake should be picked up as close to the middle of its body as possible so it can be handled safely without applying too much pressure. Most likely, the rattlesnake won't

***A mirror reflects the sun's light into the snake den.***

attempt to strike because it won't be in pain from being gripped too tightly.

You'll need a mirror to reflect the sun's light back into the den. I use a three-inch by three-inch mirror, one-fourth inch thick—the perfect size for reflecting light. A flashlight doesn't give the extra brightness that sunlight reflected off a mirror provides. I carry a six-cell flashlight to use on cloudy days, but it's too big and ineffective to bother with day to day.

Snake-proof boots are a must if you're a serious snake hunter or if you live or hunt in an area where there are rattlesnakes. My boots are all leather, handmade by the Gokey Boot Company. Snake-proof boots can be purchased for $100 to $550 a pair. They're crafted from leather, man-made materials, or a combination of both and are designed to protect the feet and legs up to the knee. If you decide snake-proof boots are too expensive, wear western or cowboy boots with your pants legs outside the boot tops. Never wear low-quarter shoes, especially tennis shoes, if you're hunting in areas of brush or high grass. A rattler's fangs go through these like a hot knife cuts through butter. You might as well be barefoot.

A spray rig plays an important part in coaxing snakes from a den. The idea is to mist a very small amount of gasoline vapor into the back of the den. To make a spray rig, solder the adjustable end of a yard sprayer to a piece of five-sixteenth inch type 'L' copper tubing. Distances into the back of dens vary. I've seen dens from three to thirty feet deep, so I have three or four

***A small amount of gasoline vapor gets the rattlesnakes to come out.***

different lengths that I keep coiled up in my Jeep until I'm ready to use them.

Don't make the tubing too long. Fasten the tubing to the sprayer valve with a flare nut or a quick coupler. The quick coupler is the easiest to use when you want to change the tubing.

Never spray more than a fine mist of gasoline. Begin at the back of the den and spray to the front. If you stop spraying midway in a den, rattlesnakes retreat deeper into the den or move to a different area.

Don't be in a hurry, even if snakes come out two or three at once. Let them exit all the way or until they're far enough out to grasp using tongs. After one or two crawl out, stop shining the sun's light back into the den and keep your movements to a minimum. Activity near the mouth of a den drives rattlesnakes farther back into the den or keeps them from coming out at all. Most dens have several exits, which provide snakes with multiple opportunities to escape.

Always carry some type of snakebite kit. I recommend the suction-type device, which comes in a kit with several suction cup sizes and some medical supplies. These can be purchased at hardware or discount stores. The suction device resembles a needle syringe; however, it works in reverse, providing *suction* when the plunger is pushed *in*.

Place the syringe cup over each fang mark and push the plunger down. When the cup fills with blood and venom, remove and empty it; then repeat the process.

There's a new device on the market, but it's not yet FDA-approved. It's a miniature stun-gun, powered by a nine-volt battery that produces 24,000 volts of current in short spurts. Using it feels like receiving a shock from a spark plug. I own one and wouldn't hesitate to use it. I have a friend who was bitten on the finger during a hunt in 1990, and the doctor who treated him used this stun-gun device on the wound. Not only did my friend's finger heal without a scar, but he also saved the mobility of his finger and hand. He showed me his hand later and said that after the doctor used the stun-gun device four or five times around the wound, the swelling stopped as well as most of the intense pain.

The suction device pulls venom out of the punctures; the electrical device neutralizes the venom. If you're bitten and use either of these devices, get to a hospital immediately! People react differently to snake venom, and you can never be too careful.

The old wives' tale about cutting an "X" over the fang marks and sucking out the venom is as dangerous as it is ineffective. Don't cut yourself or anyone else who's been bitten by a rattlesnake, and don't suck venom into your mouth. An open cut near or inside the mouth can absorb venom.

Tie a tourniquet (or use a wide rubber band) between your heart and the snakebite. If you're bitten on the hand, put the tourniquet at the wrist. If you're bitten on the foot, put the tourniquet at the ankle. Don't tie the tourniquet too tightly. You don't want to cut off circulation completely. Continue to loosen the tourniquet every few minutes until you receive medical attention.

It's very important to remain calm. Don't panic and, above all, don't drink alcohol. If you've been drinking and are bitten by a snake, the venom travels more quickly through your system because alcohol speeds up circulation. DWI (driving while intoxicated) and BWI (bitten while intoxicated) can be deadly. You can't afford either one.

# Chapter 4
# All You Want to Know and Were Afraid to Ask

Western Diamondback rattlesnakes are common as tumbleweeds in West Texas. You'll find them in town in flower beds or by the back porch, on farms under old wood or concrete slabs, in cotton fields, rat holes and where and when you least expect them.

Western Diamondbacks live long in the wild because nothing much bothers them, except the occasional human. Small rattlesnakes are eaten by roadrunners, hawks, owls and other meat eaters, but, generally, they're in charge of their surroundings.

When a female rattlesnake gives birth, she may have from five to twenty babies with a fifty percent maturity rate—excellent by nature's standards. Rattlesnakes are born alive and are not encased in shells. A few minutes after birth, their venom is toxic because they're on their own immediately. They're born with one button or the tip of a rattle.

People mistakenly believe you can tell a rattlesnake's age by counting its rattles. Each time a rattlesnake sheds its skin, it adds another rattle. If it sheds three or four times a year, it forms three or four new rattles. I've seen rattlesnakes two feet long with ten rattles. Determining a rattlesnake's age by its length is difficult. A rattlesnake five or six feet long may be ten to twelve years old. Its health and longevity depend on its environment and food supply.

If a rattlesnake strikes another rattlesnake, it seldom proves fatal to the snake receiving the bite. I've watched rattlesnakes strike each other in the head and body without injury. However, if a rattlesnake is struck where its vital organs are located—ten inches or so in front of its black and white tail bands—it will die. Not quickly, but very slowly. The main body organs—heart, kidneys, gall bladder and liver—are located in this area. The smaller the snake, the shorter the length of the organ area. A two-foot snake has an area about six inches long where its vital organs are. A rattlesnake's stomach is located near its head. If it's bitten in the stomach or the head, no injury results. Rattlesnakes are immune to their own venom.

Three or four times a year when a rattlesnake sheds its skin, it's blind until the skin comes completely off. Just before shedding, the dead skin becomes crusty and dry looking. The rattlesnake appears to perspire until its body is wet, then sucks in air and swells to twice its size. As it exhales, it crawls out of its skin through its mouth, using surrounding rocks, dried wood, or other rough surfaces to pull away the dead skin. The process takes several hours. When the last of the old skin peels off at the tip of the rattlesnake's tail, a button is added to its rattles.

Western Diamondbacks come in three colors—red, yellow, and black—the most common colors being yellow and black. Immediately after a rattlesnake sheds its skin, its color is vivid, and the diamond patterns are very prominent. You don't come across red ones often, but there are many of them around. Their color is called red, but it's really more of a dull red-orange shade.

Comparing rattlesnakes of all colors in close proximity, the difference is obvious. I once believed that their color was determined by their environment, but I've learned that their color has nothing to do with their habitat, having caught rattlesnakes of all three colors in one area.

The first time I caught a red rattlesnake was during a Roundup in the 1970s. It had just shed its skin and was a bright orange. I separated it from my other catches, certain that I had caught a rattlesnake worthy of the Guinness Book of Records! Within four or five hours, however, the snake's color had changed from bright orange to its natural dull red-orange color.

How do you tell the male rattlesnake from the female? The best explanation is that the male snake's body is uniform in size and does not taper toward its rattles. A female rattler's body narrows about twelve inches above its rattles. Otherwise, they appear to be identical.

A rattlesnake has fourteen fangs in the roof of its mouth, seven on each side. People have bragged that they've "de-fanged" a rattlesnake, but the only way to do that would be to cut out all the fangs. Even then, the rattlesnake would grow more fangs if it survived the surgery, which is unlikely.

The front two fangs inject venom when the rattlesnake strikes. Two fangs located just behind these inject venom if the front fangs are lost or damaged. As front fangs are shed, a second set moves forward to replace them, and new fangs begin growing in the back. I've caught rattlesnakes with two fangs on one side of the mouth. If this snake were to strike, one of the double fangs would be left sticking in the bite mark.

A rattlesnake's lower jaw is fitted with small teeth that taper backward, enabling them to hold small prey, such as rats, but these teeth don't inject venom. If their prey is large, rattlesnakes strike and immediately recoil to strike again.

Rattlesnakes normally strike outward. The only rattlesnake I've caught that struck straight up was a red rattler that apparently had been injured previously. This rattlesnake came close to striking me several times. I had placed it, along with other snakes I had caught, into a wooden box about fourteen inches deep.

***A Rattlesnake has a total of fourteen fangs. Behind these first two are smaller ones. Each time it sheds its skin it loses the front two and two more move up.***

Every time I opened the lid, this rattlesnake struck at my hand, coming up about six inches out of the box opening.

Being a fast learner, I moved it to a separate box. I took the snake to Bill Ransberger's snake-handling demonstration enclosure at the Roundup and told him I had something special for him. He lifted Crazy Red from my snake box and placed him on the table in the enclosure. For a few minutes, everything was fine—just an unusual red rattlesnake on display. Then Crazy Red lunged from the table at Bill and fell to the floor. Onlookers gasped loudly, but Bill wasn't aware of the near miss. The second time Crazy Red struck at Bill from the floor, it almost hit him on his rear end. Bill, the ultimate showman, didn't display any undue concern, but Crazy Red was soon out of Bill's demonstration enclosure and on its way to the cook shack where it was killed, skinned, breaded, and fried.

A rare instance of a rattlesnake striking its full length occurred when a six-foot one struck Bill during a roundup in Andrews. The rattlesnake was coiled in the corner of the enclosure where Bill was giving a demonstration. Bill looked away for a second, and the rattlesnake lunged up and out its full length, striking him with both fangs, injecting a large amount of venom into a major vein on Bill's left hand.

This extremely dangerous and painful bite resulted in Bill's extended stay in an Andrews' hospital. The doctor had to slice open Bill's hand, arm, and armpit to extract the venom.

While Bill was in the hospital, I called to check on him. He told me that after he recovered he intended to keep the seventy-two metal stitches the doctor had used to close the wound.

Later, I asked what happened to the rattlesnake that bit him. He grinned and said, "I ate him—French-fried!"

When a rattlesnake strikes, either it feels threatened, is killing prey for food, or is blind while shedding its skin.

Rattlesnake's eyes have no lids—just a plastic-like coating—and never close, even while the rattlesnake is sleeping.

A rattlesnake eats its prey alive, or almost alive, and swallows it whole. Their jaws are hinged so they can swallow something twice the size of their mouth, working it slowly down using the strong muscles in their body. When they strike a rabbit or a rat, they don't have to hold on. Their victim goes into shock from the effects of the venom and runs only a few yards before collapsing. The prey stays where it falls, going deeper into shock, and the rattlesnake trails it to that spot.

The black, forked tongue, where its sense of smell originates, is common to all snakes, even nonpoisonous ones. It's very sensitive and leads the rattlesnake to its prey. Rattlesnakes can't hear or smell well, but they are experts at picking up vibrations through the holes located between their eyes.

However, I've caught rattlesnakes sleeping soundly coiled in front of their dens by carefully avoiding dry sticks or loose rocks. Because I didn't make noise, they didn't know I was there.

***It takes at least twenty-five rattlesnakes to produce this much venom. Snakes at the Sweetwater Rattlesnake Roundup aren't milked if they are less than twenty-four inches long.***

A rattlesnake's digestive system is amazing—capable of dissolving tissue, bones, feathers, everything. Hours later, from a small opening near its tail, the snake secretes small black pieces of matter surrounded by yellowish albumen, similar to chicken's egg white.

Their digestive juices and venom are produced by enzymes in their body. If it were possible to "milk" a rattlesnake dry, it could replenish its venom supply in eighteen to twenty-four hours. Their venom is the color and consistency of orange juice, and twenty-five to thirty large rattlesnakes must be milked to produce one c.c. of venom.

*More than three hundred rattlesnakes were captured from this one den near Sweetwater.*

Sweetwater Jaycees donate the highly-toxic venom they collect during the Roundup to medical laboratories for use in cancer and other disease research. It's ironic that rattlesnake venom is used for lifesaving purposes while their bite damages and destroys cell tissue.

Rattlesnakes drink water and swim gracefully, though it's not their favorite activity. Their normal body temperature is fifty-five degrees. Rattlesnakes stay close to their dens beginning in October and November when the weather begins to stay cool for longer periods. During these cooler months, rattlesnakes enjoy basking in the sun's warmth near the mouth of their dens. If they're out of their dens at night when temperatures drop low or near freezing, chances are they won't survive.

You won't see them out in the sun at noon on hot West Texas summer days either. If they get too hot, they won't survive.

Snakes can't dig their own dens, so they move in with rats, mice, rabbits and skunks. When a rattlesnake is ready to hibernate, it isn't particular about its roommates for the winter. In their dens, rattlers snooze through the winter months, beginning to venture out again in late February or early March as daytime temperatures warm into the mid-60 and mid-70 degree ranges.

Rattlesnakes avoid open spaces and prefer areas of brush, fallen trees or high dense weeds. They're homebodies and rarely travel more than a one-mile radius from where they're born.

On West Texas summer nights, it's not uncommon to see them on highways soaking up the warmth of the pavement. It's against

Texas Parks and Wildlife laws to hunt rattlesnakes at night by picking them up off the pavement. Several years ago, the Texas Legislature passed a law that you must carry a State-issued rattlesnake hunting license to have more than ten rattlesnakes in your possession at any time.

# Chapter 5
# You Have to Start Somewhere

In 1960, when I became a member of the Jaycees (Junior Chamber of Commerce) in Sweetwater, I had no idea it would result in my active participation in the World's Largest Rattlesnake Roundup for so many years. Or that my association with rattlesnakes would open so many doors of opportunity.

The Sweetwater Jaycees originated the Roundup in 1958 as a fund-raising project. The first Roundup was held in the Nolan County Coliseum, and my job was collecting tickets at the door. It wasn't a demanding job because attendance wasn't high—mostly local residents who came to watch the young

***Tom Wideman, as president of the Sweetwater Jaycees in 1965, weighing and measuring a rattlesnake.***

Jaycees participate in their crazy once-a-year ways and means project.

I had a strong fear of rattlesnakes and promised myself that I would *never* hunt or handle the dangerous reptiles. In the forty-eight years since, life has taught me many things—chief among them, never say never.

When the three-day event in 1958 ended, I viewed rattlesnakes with much less apprehension. In 1959, I assisted in the weigh-in pit, writing daily poundage totals on a large blackboard. Two years later I was in charge of the weighing and measuring pit. I had come a long way from taking tickets at the door.

Every year the Roundup grew in scope and size, and my participation did the same. In 1962, fellow Jaycee Bill Ransberger and I co-chaired the Roundup. Back then, it was a challenge to accept chairmanship because of the planning and time it required. Looking back, however, the first Roundups seem uncomplicated compared to the time, effort, and planning involved in chairing the event now. In 1962, we were a long way from producing the well-organized event that unfolds the second weekend in March each year in Sweetwater, drawing an estimated 35,000 to 40,000 visitors from all over the United States and several foreign countries.

I returned to the weigh-in pit in 1963 and kept that responsibility until 1965, when I was elected president of the Sweetwater Jaycees. Working in the weigh-in pit was—and is—a dangerous job. Rattlesnakes are generally brought in directly from their dens and haven't been milked, so they're full of venom. Most of them are hot, meaning they're extremely agitated—first at being captured and second at being transported in a container filled with dozens of other ill-tempered rattlesnakes. When containers are opened, rattlesnakes inside strike wildly at anyone and anything within reach and are truly armed and dangerous.

The Sweetwater Roundup used to feature a Smallest Snake competition, in addition to the Largest Snake trophy. I was in charge of weighing and measuring snakes to determine potential winners in both categories. Small rattlesnakes are especially dangerous because their venom is concentrated and toxic within minutes of their live birth. I compare small rattlesnakes, which are extremely difficult to grasp with bare hands, to rubber bands. If you grip them too tightly, you can kill them. If you grip them too loosely, they stretch and pull from your fingers.

In 1964, a rattlesnake less than ten inches long pulled from my grasp and nicked the nail of my left-hand ring finger with one fang as I was stretching it on a table for measurement. Luckily, it didn't inject venom as its fang scratched my fingernail. Snake handlers call this a dry bite.

Fellow Jaycees who witnessed the incident urged me to go to the hospital, but I wouldn't be persuaded. Reflecting on the incident now, I probably should have let a doctor take a look, but I was younger then and thought myself bullet-proof and invisible. Following the incident, Jerry Ransberger, Bill's son, made a spe-

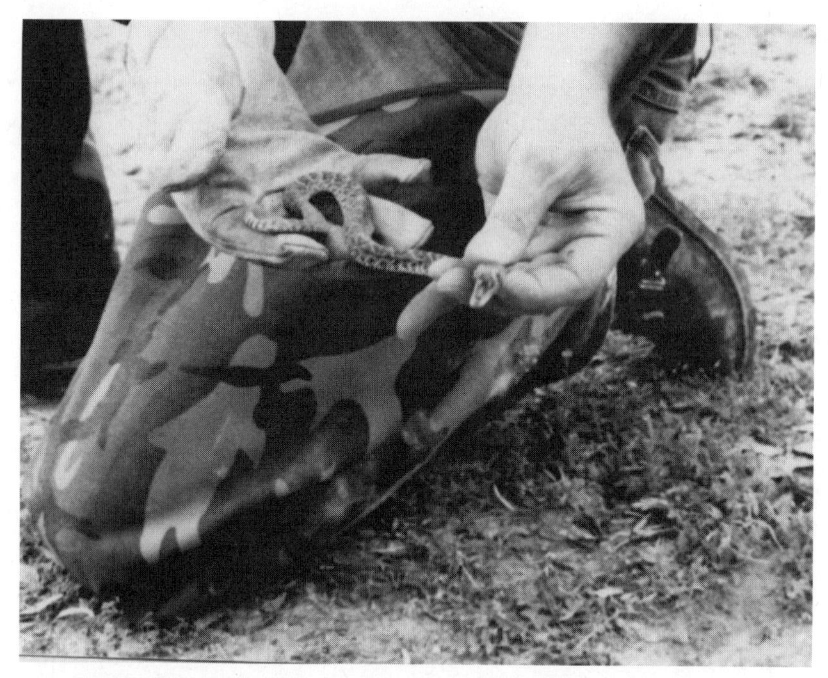

***Too small to handle: Tom received a 'dry bite' from a small rattler in 1964. The small snakes are too small to get a good grip on, which can be very dangerous for the handler.***

cial pair of snake tongs, which I still have and use when handling small snakes.

The Smallest Snake contest was discontinued in 1965 for ecological reasons and the risks associated with handling them. Bill ended his snake-handling career after being bitten forty-one times, along with many near misses.

It took only one dry bite to get me out of the Coliseum and into the countryside.

# Chapter 6
# The Greenhorn

My first encounter with rattlesnakes in their natural habitat was in 1966, the year I invited myself on a rattlesnake hunt with Ikie Wilson and Finis Kennon, who had been hunting partners for many years.

It seems like I've always known Ikie Wilson and appreciated his friendship and admired his hunting skills. During my growing up years in Sweetwater, Finis and Opal Kennon treated me as a member of their family. I attended school with their son, Don, and after high school, he and I enrolled at Sul Ross State University in Alpine, both of us graduating in 1959. Don joined the Marines after graduation and was a lieutenant colonel when he died a hero in Vietnam.

Until that first hunting trip with Ikie and Finis, my only contact with rattlesnakes had been during Roundup activities in the

Nolan County Coliseum, a totally controlled atmosphere. I didn't know how to hunt rattlesnakes in the wild, but Ikie and Finis did, and they taught me an extremely valuable lesson—sometimes only a sharp eye can keep you out of trouble.

The day I went hunting with Ikie and Finis, we traveled to a ranch near Sweetwater, my first visit to a rattlesnake den. I didn't know how to locate rattlesnakes, and I sure didn't know what to take along on a hunt. Although I had been told we would be away from town all day, I took nothing to eat or drink. I didn't even take any equipment, planning to catch them bare-handed, I suppose.

Ikie and Finis teased me about being such a greenhorn, especially when it came time to eat. After some kidding, they shared their lunches with me. With experience gained through the years, I've done the same thing for greenhorns I've taken out hunting.

When we arrived at the first rattlesnake den, we discovered that it opened on a huge flat rock perched on top of a hill, facing east. I'll never forget the excitement I felt as we walked up to the den. Rattlesnakes lay coiled everywhere!

We had brought three forty-gallon metal cans, along with two snake tongs, which Ikie had made by hand. Using these tongs, Finis and Ikie began tossing the rattlesnakes back to me so quickly that even with my snake hook I couldn't put them in the metal cans fast enough to keep pace.

My adrenaline started pumping! I shouted at them for some help. Finis came to my rescue, laughing, and said, "I thought you were a rattlesnake hunter!" I replied, "Right now, I don't know what the heck I'm doing." They both laughed, and I got a feeling that the whole routine had been to initiate the greenhorn rattlesnake hunter into the "club."

We filled the three forty-gallon cans so full that rattlesnakes were slithering out over the tops as quickly as we put them in. Ikie said that he had some burlap sacks in his International Travel-All and asked me to go get them while he and Finis put lids on the cans to keep the rattlesnakes from escaping.

The burlap sacks sounded like a good idea until I found out that one of us had to hold the bags open while the other two put rattlesnakes inside. My blood pressure climbed out of sight again when they "elected" me to hold the sacks open.

Occasionally, a rattlesnake's sharp fangs would get caught on the sack, and we'd have to free it before continuing. After an hour or so, we had caught more than two hundred snakes. It was Saturday afternoon, and the Roundup closed at five o'clock.

We were running out of time to get back to town, but the rattlesnakes kept coming out of the den, four or five at a time. The den was about twenty feet deep, and by using a mirror to reflect the sunlight, we could see hundreds of them still in the back of the den. But, it was almost four, and Ikie decided we had better put our catch in the Travel-All and head back to the Coliseum.

I can't remember how many rattlesnakes we turned in that day, but it was more than three hundred pounds. I wouldn't be surprised if we didn't leave that many more rattlesnakes back in the den.

West Texans are big kidders, of course, especially around newcomers. That same afternoon, a man in the Coliseum asked Ikie where he could catch some rattlesnakes to take back home with him. Ikie asked where he lived, and the man replied, "Illinois." With a serious expression, Ikie said, "Get some snake tongs and a burlap bag. Go out on the highway and look on the shady side of each fence post you pass. Rattlesnakes like to lie in the shade."

The man walked away, and we thought nothing more of the conversation. The next day as we drove toward a ranch where we were going to hunt, we passed a car parked along side the highway. Ikie said, "Hey, look, that's the guy from Illinois." Sure enough, the fellow was looking on the shady side of each fence post, keeping his snake tongs and burlap bag at the ready.

I'll always be indebted to Ikie Wilson and Finis Kennon for sharing their experience and rattlesnake hunting skills with me, along with their West Texas humor, even though they did almost scare the wits out of me that first trip.

# Chapter 7
# Ol' Tuffy

Do you believe in love at first sight? It happened to me in 1970 when I saw a 1946 CJ Willys Jeep parked on Otis Wafer's used car lot in Sweetwater.

The Jeep was thirty-four years "young" and a wreck, but I didn't care. It was a World War II vintage Jeep with flat fenders and flat hood. The engine was original factory—flathead, four-cylinder, and the cast iron cylinder head top was embossed "Willys." The Jeep had been painted many times during its life—orange, red, green, white, and some off-color black/gray, probably primer coat for the other colors, and all these colors showed around the frame when you pushed in the clutch or used the brake. It was similar to looking at a rock cliff where you can read ages past in the earth's crust. Surprisingly, Otis didn't offer to *pay me* to take it off his lot.

***Tom has driven Ol' Tuffy on snake hunts for thirty-five years.***

I towed my "new" Jeep to Carmichael Ford where Eddie Neitzler was service manager and his brother, Leo, head mechanic —two of the best in the business. The shop crew helped me push the Jeep into the service bay, looking at the Jeep and me in disbelief. I asked Eddie to take a look at the engine, and a couple of weeks later he called and said we'd better talk.

Despite his less than optimistic prognosis, I gave Eddie the go-ahead to do whatever it took to get the Jeep running. Responsibility for putting the chassis back together fell to me and Melvin Guelker, a genius at fashioning metal into something beautiful and useful.

It took us almost a year to get the Jeep in shape to paint. After years of restoration, its top speed is forty miles an hour, but it's geared very low and will take you anywhere you're brave enough to go. The Willys Jeep, which I later named Ol' Tuffy, has never let me down, and we've been to some rough and dangerous places together.

I warehoused the Jeep and my '57 Thunderbird in a building on Lamar Street. The building was thirty feet by sixty feet and ten feet tall, constructed of native rock—a solid building by any standards. However, in April 1986 it was put to the test. I was at my Lake Sweetwater cabin, where heavy rain and hail fell, accompanied by forty to sixty mph winds. Then, everything got deathly still. I looked toward Sweetwater, and the sky was a gray-green color.

I got into my pickup and headed toward town. A tornado had ripped through Sweetwater's south side beginning at Lamar Street along Interstate 20. National Guard and state and local law enforcement people were everywhere. No one was allowed access to the area, but I contacted the Nolan County sheriff, who confirmed that my warehouse had taken a direct hit and authorized my access to Lamar Street, which looked like a war zone. Trees, houses, cars, clothes, and furniture were shredded and scattered everywhere.

When I reached my warehouse, I found it crushed, not by the tornado, but by the weight of the roof of the business next door. The roof had been ripped off and tossed onto my warehouse, collapsing it except for one sixty-foot wall, which was leaning at a sixty-degree angle and resting on something. I couldn't get in, but I could see that the contents had sustained heavy damage.

City-owned heavy equipment was being used elsewhere, but a city employee promised that they would come back the next day and pull the wall over so I could see what could be salvaged. Early the next morning I was at the warehouse, and around ten o'clock a city crew arrived with a front-end loader. They attached a steel cable to two steel posts holding the rock sides of the warehouse in place. After three hours lifting the roof off my building, the crew pulled the wall away.

When it hit the ground, dirt and dust flew everywhere, and after it settled, I could see that almost everything in the warehouse was ruined. The '57 T-bird was totaled inside and out. The tires were blown, and there wasn't an area ten inches square on the chassis that wasn't dented. Broken glass lay everywhere, and the T-Bird's roof had a huge hole in it.

The Jeep, however, which had been covered only with a cotton dust cover that had blown away, had been holding up the ten-foot tall, sixty-foot long wall. Otherwise, the wall would have been on top of the T-bird as well. I was amazed at the Jeep's condition. The windshield was down, but the glasses were unbroken. There wasn't much dirt or rock on the seats. None of the tires were flat. No scratches—only a golf ball-sized dent on the edge of the left fender with a scratch deep enough to reveal some red and orange paint from years past.

It was the city crew's turn to be amazed when I got into the Jeep, turned the key, mashed the starter button, and the motor started. I backed the Jeep over the rubble out into the alley. As I backed it out, someone said, "Lord, that old thing is tough!"

That's how my Jeep got its name. I had a plaque engraved for the dashboard—Ol' Tuffy.

Ol' Tuffy has been seen by millions of people in a National Geographic Explorer segment, filmed in 1991 to document that year's Rattlesnake Roundup and still aired from time to time. National Geographic Explorer has combined the Sweetwater Roundup footage with segments featuring the world famous Chinese Acrobat team, an African Safari, and Search for the Titanic. Several airlines show the Roundup segment in their Travel Guide before in-flight films.

In 2005, when I marked my forty-fifth year of involvement in the Rattlesnake Roundup, Ol' Tuffy celebrated its thirty-fourth year with me.

You're famous, Ol' Tuffy, and I still love you!

# Chapter 8
# West Texas Folks

Paul and Margaret Newman Hill have been close friends for as long as I can remember and have graciously opened their ranches to me for snake hunting. Without their encouragement, I doubt I would have stayed involved in the Sweetwater Roundup for so long.

Paul and Margaret are the type folks who give West Texas its good name. Their families are part of the heritage and history of Nolan County. Margaret always makes you feel welcome and comfortable, and her smile lights up a room. Paul's sense of humor was as dry as the desert, and he loved to kid. He enjoyed teasing me through the years. When he passed away in September 1999, I felt as if I had lost a member of my family.

The Hills' Palava Catfish Ranch north of Sweetwater is a prime example of rugged West Texas country. It's big, rough, and difficult to navigate. Landmarks on large West Texas ranches change on a daily basis, and the Catfish Ranch presents a different face each time we go rattlesnake hunting.

Paul was an excellent guide with an uncanny sense of direction, and he knew every inch of the terrain on his ranches. I, on the other hand, have been lost on the Catfish so often that it's downright embarrassing. Of course, Paul knew exactly where we were at all times and delighted in seeing me bumfuzzled as to north, south, east, or west.

After snake hunting on the Catfish for seventeen years, there was one den, lovingly called The Flea Hole that I could never locate without Paul's help. After I had knocked down trees and brush and had lost both my senses of direction and humor, Paul would quietly say, "Well, it's right over here, Tommy." Although he never said as much, I could tell by the twinkle in his eye that Paul was delighted to have to come to my rescue.

One year, he told me to follow him to the north end of a sprawling mass of rugged terrain to a den he had never shown me. I was driving Ol' Tuffy, and Paul was driving his pickup. We drove four or five miles back into the middle of a pasture covered in brush, rocks, and thick undergrowth. As we approached the den, Paul said that he had seen twenty or more rattlesnakes out by the den's mouth, which was near a sink, or depression in the

***Paul Hill knew where to find rattlesnakes on his Palava Catfish Ranch north of Sweetwater.***

ground, about fifteen feet across and three feet deep. The sink was covered in green wild rye winter grass.

When we reached the den, rattlesnakes littered the ground. They ranged in size from three to four feet, with a couple of five-foot snakes thrown in for good measure. We picked up twenty or more and placed them in our snake boxes.

Paul, in his usual dry, matter-of-fact style, told me that he had seen a really large rattlesnake in the bottom of the sink the day before. As always, when hunting, I was wearing Gokey snake-proof boots that reach almost to my knees. I stepped down the three feet to the bottom of the sink while Paul watched from above. My right foot no sooner hit the ground than a six-foot rattlesnake came out of hiding and struck the top of my boot. I had stepped almost right on top of him. The sudden strike shook me up, and I believe that I jumped straight up three feet out of the sink and landed next to Paul.

He began laughing, and said, "Boy, you're getting a little antsy, aren't you, Tommy?" I replied, "You think so? Well, you get down in there and see what you do!" He grinned and said, "That's not my job. That's your job!"

After I settled down, we both enjoyed a good laugh. I told Paul that I had to catch that snake, so I got my sprayer, mirror, and snake tongs from Ol' Tuffy and stepped back down into the sink—fortunately, not on top of a rattlesnake this time.

The sink had holes 360 degrees around the bottom, and I could see several more rattlesnakes reflected in the light of my mirror. Paul and I planned our strategy before using the gas vapors because we figured when the den was sprayed, rattlesnakes would come out in all directions, all at once.

Paul agreed to put them in the snake box as I caught and lifted them up out of the sink. I sprayed the area, and two and three snakes at a time emerged from the den. From this one den, in just a couple of hours, we caught seventy-five large rattlesnakes.

Every time I hunted with Paul, we shared new adventures and collected more snake tales.

# Chapter 9
# Bonus Baby

Donnie Anderson, the Texas Tech football star, was among the first "bonus babies" to gain fame and fortune with the Green Bay Packers during the years they were coached by Vince Lombardi.

Donnie attended the Rattlesnake Roundup in 1966 to serve as a judge for the Rattlesnake Queen contest. My friend Ed Aiken and I had the pleasure of taking him on a rattlesnake hunt on Ed's ranch one afternoon.

His presence caused quite a stir among Roundup attendees. A large group at the Coliseum wanted to accompany us on the hunt to be photographed with him. I enjoy being around people, but the size of this crowd was more than I had bargained for. We didn't want to disappoint anyone, so Dewey Nelms, then ranking officer of Sweetwater's 36th Infantry National Guard Unit, agreed

***Football star Donnie Anderson, left, was a good sport on his first rattlesnake hunt.***

to let us use several of the Guard's two-and-a-half-ton trucks to haul the crowd of hunt participants, onlookers, and press people.

We were quite a sight as our caravan drove to Ed's ranch north of Sweetwater. I had been hunting all week with reporters and feature writers from the Hughes Tool Company. Miles Knape, editor of *Hughes Rigway* magazine, was researching a feature story about the hunt and was one of several photographers on the field trip hoping to get a photo of Donnie Anderson holding a rattlesnake with tongs.

When we got to Ed's ranch, we drove to a hill covered with large rocks. Ed had told me that the day before he had seen several large rattlesnakes in this location. I cautioned Ed that we had a potentially dangerous situation because people in the large crowd were behaving as if we were at a petting zoo. People were literally falling out of the various trucks, pickups, and cars in an attempt to watch Donnie's every move.

Donnie was naturally a bit apprehensive and didn't venture far from Ed and me. He was dressed for his Rattlesnake Queen contest judging duties in a light gray silk suit with a sport shirt unbuttoned at the collar, and he looked like a million dollars.

While press people and onlookers crowded around Donnie to ask questions, Ed and I quietly made a brief tour of the area. Using mirrors to reflect the bright sunlight, we found some large rattlesnakes in a den tucked under an imposing rock ledge. When the large crowd realized what Ed and I were doing, they hurriedly gathered around us. It sounded like a herd of stampeding cattle.

We sprayed some gasoline vapors under the ledge to encourage the rattlesnakes to come out so Donnie could catch one for the benefit of the reporters, camera crews, and spectators. With Donnie standing behind me on what turned out to be an unstable ledge of rocks, I used sunlight reflected off my mirror to spot a rattlesnake near the front of the den.

I spotted a nonpoisonous five-foot coach whip snake. I didn't want Donnie to hurt the coach whip, knowing that if he grabbed it with the tongs, he might press too tightly in the excitement of the moment. The coach whip slowly crawled a foot or so from the mouth of the den and then shot straight out. The large crowd that had gathered around Donnie, Ed, and me, took one look and ran. Suddenly, the three of us were alone.

I grabbed the coach whip with my hand by its tail, and it rolled up over my hand, hitting me in the chest. Donnie thought that I had caught a rattlesnake bare-handed and that it had bitten me on the chest. In his excitement, he fell backward on the rocks—silk suit and all.

By then, the only creatures on the ledge were Ed, the coach whip, and me. I held the coach whip in my hands until everyone settled down. Donnie picked himself up, brushed dust and dirt from his suit, and muttered, "I'd rather take on the biggest lineman in the league than do this!"

The crowd relaxed a bit after that and walked closer to the den. Donnie was a great sport because he walked back to the den with the snake tongs and said, "Let's try again."

This time, reflected in the light from my mirror, I saw a four-foot rattlesnake easing its way toward the mouth of the den and pointed him out to Donnie. When the rattlesnake came out far enough to reach, I told Donnie to grab it about a foot back from its head and to try not to cut it in half with the tongs. He grinned and said, "We'll see."

I told him to make his first try his best because if he missed, the rattlesnake would head back into the den and wouldn't come out again any time soon.

Donnie caught the snake on his first attempt.

Cameras clicked everywhere, while press people shouted out directions. "Hold the snake this way… that way… look over here… smile… be serious… pretend you're having fun." Patiently, he complied with each photographer's request.

Someone asked Donnie if he thought Coach Lombardi would be impressed by the photograph when he saw it in a newspaper. He laughed and said, "What do you think?"

Donnie Anderson's good sportsmanship made the hunt special for all of us. We stayed at Ed's ranch most of the day and caught more than thirty pounds of rattlesnakes—a respectable catch when you consider that there were only two people in the crowd who were experienced rattlesnake hunters.

I'm amazed that someone in that huge crowd didn't step on a snake during that wild afternoon.

## Chapter 10

# One for the Road

I met Craig Eckermeyer and his wife at the Roundup in 1964, after Craig learned of the Sweetwater event from a nationally televised news broadcast.

Craig, a Standard Brands Corporation executive, and his wife lived in Camden, New Jersey. An amateur herpetologist, he kept several poisonous and nonpoisonous snakes in glass cases in the basement of his home.

When we were introduced at the Nolan County Coliseum, Craig had already been out hunting rattlesnakes on his own. He hadn't caught any and asked if he could hunt with my group, which already included ten hunters and several press people. I told him that in spite of the large group we would make room for him and his wife. Craig was excited to be part of our hunting expedition and hesitated just long enough to tell several of his East Coast friends at the Coliseum why they were being left behind.

The first den we scouted was north of Sweetwater, and we saw several rattlesnakes sunning themselves in front of their den. Craig reached the den first but, because of noise and movement generated in his excitement, the snakes quickly crawled back into the mouth of the den. I asked Craig to slow down a bit and to watch until the first rattlesnake had been caught.

He was embarrassed that his actions had caused the rattlers to disappear and was afraid that they were gone for good. I reassured him that the rattlesnakes would come back out after we sprayed some gasoline vapor mist into the den.

We hunted all day and didn't head back to Sweetwater until almost sundown. When our group arrived at the Coliseum, we began unloading the day's catch. I told Craig that we would be going out again the next day, Sunday, and that he was welcome to hunt with us. He couldn't go, however, because the Eckermeyers and their friends had to begin their long drive back to New Jersey.

As I lifted the last box of rattlesnakes from the bed of my pickup, Craig asked for a "favor." He wanted to take three or four large rattlesnakes back with him to New Jersey because he didn't have any Western Diamondbacks in his collection. He reminded me that he was a herpetologist and assured me that he knew how to handle snakes.

Against my better judgment, I agreed, rationalizing that he and his wife had traveled a long distance to attend the Roundup and he was experienced with snakes.

The Eckermeyers had their own snake box, so he picked three or four of the rattlesnakes we had caught that day, as excited as a kid with a new toy, even though I still had my reservations. A couple of days later I received a call from Craig letting me know that they had made it home safely and that he really appreciated having the Western Diamondbacks.

I didn't hear from Craig again until the following March when he and his wife returned to Sweetwater for the 1965 Roundup. When I recognized them in the crowd at the Coliseum, Craig reached out to shake my hand, and I noticed that the thumb on his right hand was red and stiff and covered with scar tissue. It looked awful. I suspected that he had been bitten, so I asked what had happened.

He said, "I was using my snake hook to move the largest rattlesnake to another glass case. It got on top of the snake hook shaft and struck straight up the shaft, hitting the top of my thumb with both fangs." He managed to get the rattlesnake back into the glass case and then went to the hospital.

The scarring I noticed resulted from having his thumb cut to extract venom and reduce swelling. The thumb joint was so stiff that he couldn't bend his thumb—the flexibility completely gone. I was grateful that Craig wasn't injured even more severely.

He kept the rattlesnake that bit him. In fact, he kept all of the rattlesnakes he had taken from the Roundup the previous year.

Craig and his wife returned for the Roundup every year until 1967, but I haven't heard from them since. Craig's accident resulted in one of my strictest policies: I don't allow anyone to take a live rattlesnake from the Coliseum. Today if you were to ask me for "one for the road," my answer would be, "Absolutely not!"

# Chapter 11
# A Real Georgia Peach

Not long after she won the Miss Georgia title, Amanda Smith attended the Sweetwater Rattlesnake Roundup in 1985 to serve as a judge for the Queen contest. Terry Staton, a Sweetwater native and Amanda's close friend, accompanied her, and Terry's family served as host for Amanda and her Miss USA Pageant chaperones.

A photo opportunity had been arranged for her to hunt a pre-baited den, but to the chagrin of her chaperones and the surprise of practically everyone, she asked to take her chances with a "for real" rattlesnake hunt. I was told on Thursday morning that she and her entourage wanted to go hunting with the group I was leading. I was happy to oblige.

Amanda was excited about going to a rugged West Texas area to hunt. She dressed in a colorful Western shirt tucked into blue jeans, snake-proof boots, and a Roundup celebrity cap. When our caravan of hunters and press people arrived at the Palava Catfish Ranch north of Sweetwater, Amanda was the immediate center of attention.

However, she was more interested in catching a live rattlesnake than in posing for the cameras. My hunting equipment, including snake tongs, mirrors, sprayer, and snake box loaded in Ol' Tuffy, was already at the Ranch. Ol' Tuffy doesn't have a top, and I had tied the windshield down in front. I told Amanda that she could ride in an air-conditioned pickup if she preferred because riding in Ol' Tuffy would be rough and dusty.

She gave me a dazzling Georgia smile and said, "I'm with you and Ol' Tuffy for the day. The dust and cedar brush are part of this adventure, and I want to make the most of it."

Amanda asked her friend, Terry, and her chaperones to catch a ride in a pickup because she wanted to ride in my Jeep. Mercy, what a deal for me and Ol' Tuffy!

We worked a couple of places before we came upon several large rattlesnakes sunning out in front of a den. When Amanda saw them, she stepped quietly out of Ol' Tuffy and asked about the best way to catch them with the snake tongs. She wasn't the least bit apprehensive, although I can't say the same for her chaperones.

*Miss Georgia Amanda Smith with her hunting guide, Tom Wideman.*

I told her to pin the rattlesnake with the tongs close to the middle of its body, if possible. As she touched one of the rattlesnakes, it jerked away, sliding quickly back into the mouth of the den. She was disappointed, but I reassured her, saying, "Don't worry! We'll spray the den and more rattlesnakes will come out." We misted the back of the den with gasoline vapor, waited a few minutes, and two rattlesnakes emerged simultaneously.

Amanda captured the first one, and I got the second. She was very excited and pleased. Cameras clicked from all directions.

We stayed at the ranch all day and caught about sixty pounds of rattlesnakes. Amanda was really into snake hunting by mid-afternoon and wandered away from the crowd several times in an attempt to find other dens on her own. She was one of the most natural people—male or female—I've had the good fortune to take out on a hunt, a lovely young woman with a genuine love of the outdoors.

# Chapter 12
# Arkansas Tunnel Rats

Don Bunch and Bill Fink, who became two of my favorite snake hunting buddies, traveled from Fayetteville, Arkansas, for six consecutive years to take part in the three-day Sweetwater Rattlesnake Roundup.

The two men owned and operated Universal Tong Company in Fayetteville, and their first visit to the Roundup was prompted by their interest in promoting and selling the tongs they produced to snake hunters.

Don worked as a mechanical engineer with Reynolds Aluminum, and Bill owned a retail pet shop. The two became partners in the Universal Tong Company as a sideline business. Both were avid outdoorsmen, very comfortable in their home state of Arkansas; however, the barren expanse of West Texas was new to them, as was the sport of rattlesnake hunting.

On their first visit to the Roundup, they were introduced to my friend Trace Fomby, who invited them to accompany him on a rattlesnake hunt near the site of what used to be the town of Eskota, now a West Texas ghost town. The three men hunted diligently all day but caught only a nonpoisonous coach whip snake and a pack rat.

Because snake hunting was new to Bill and Don, Trace told them that he knew a guy they needed to meet—me, as it turned out. It was the beginning of a great friendship.

I'll never forget the first time I saw Bill Fink. He was wearing blue jeans, motorcycle racing boots, and a Rattlesnake Roundup tee shirt with a red banana tied around his neck. He had a full, gray beard and wore a large cowboy hat with a rattlesnake head mounted on the brim. Bill looked like a member of Hell's Angels who had lost his way.

Don Bunch was also dressed in blue jeans and a tee shirt. He was wearing mountain climbing boots and bright red *Budweiser* suspenders, and he sported a dark beard reaching almost to his stomach. The cap he was wearing read "Arkansas-The Natural State."

The way Don and Bill were dressed stands out in my memory even though I've seen a variety of clothing combinations during my forty-five years at the Roundup. People from all walks of life

*Arkansas 'tunnel rats' Don Bunch, left, and Bill Fink with Ol' Tuffy.*

attend the event every year—doctors, lawyers, secretaries, engineers, bankers, data processors, ranchers, homemakers, business executives—each dressed to express his or her unique style at this most unusual sporting event.

Don and Bill earned their nickname, "Arkansas Tunnel Rats," that first weekend. While we were hunting, we came across an area covered with holes and gypsum sinks—deep depressions in the ground. One large hole had an entrance about three feet in diameter. I had hunted there before but had never gone further than five or six feet back inside the hole.

The gypsum surrounding the large hole isn't rock hard. It's soft and falls apart when touched or bumped. Even when you direct the beam of a flashlight back into the hole, you can't see very far because of its many turns and crannies. During numerous visits to the area, I had seen only one lone rattlesnake out in front of the hole and had never taken time to investigate it thoroughly. Besides, I believed that tunneling in was too dangerous.

Bill Fink, on the other hand, was a person seemingly undeterred by danger when it came to the out-of-doors. He thought he *had* to go into the cave although I told him that it wasn't safe due to the gypsum rock's instability. Bill didn't care—he was going to the back of the cave or bust.

Don and Bill had brought a hundred-foot extension cord and a spotlight on our expedition. Don hooked the cord to the truck's battery, and Bill and another hunter in our group, Shawn, began their crawl into the cave.

I was extremely apprehensive. After taking countless people on hunts without anyone being hurt or snake bitten, I wasn't ready to ruin my record. In addition to my concern about rattlesnakes they might encounter during their crawl, the danger of the cave walls collapsing weighed heavily on my mind. Even Don was concerned for his friend's safety, though he kept these thoughts to himself, knowing that I was worried enough for both of us.

About forty-five minutes later, almost all one hundred feet of the extension cord had been pulled into the cave, and we could no longer hear Bill and Shawn's voices. Then, the cord stopped moving. By this time, I was positive that they were in trouble.

I didn't realize that Bill and Shawn had reached the back of the cave and were crawling back out. When they were back within hearing distance, we called their names. Within minutes, both appeared, covered with gypsum dust.

"That was great!" Bill exclaimed. "There must have been some coyotes living in there at one time!" Those tenants apparently had abandoned the cave, because Bill and Shawn found only some animal bones and broken glass.

I didn't say much to Bill then because I was relieved to have both hunters safely out of the cave. Later, I told Bill and Shawn that there would be no more cave exploring while I was in charge of the hunting expedition—end of conversation!

Bill Fink was quite a man—a little wild maybe, but obviously not afraid to test the boundaries of his courage or the laws of nature. Unfortunately, several years later he lost his life while on a camping trip at Beaver Lake in Arkansas with his buddy, Don Bunch. Bill drowned while attempting to swim across the lake against a raging tide of frigid water.

Don and I kept in touch, but he stopped coming to the Roundup. The death of his best friend changed his life.

# Chapter 13
# Snakes Without a Compass

One of the most capable people it's been my good fortune to meet and hunt with is Karen Boggs. Karen, who was the first female member of the Sweetwater Jaycees, is petite and pretty and also very athletic. She doesn't mind hunting in the roughest areas, catching rattlesnakes as well as or better than many men I've hunted with.

I was introduced to Karen during the 1985 Roundup and had no idea what a sportswoman she was. Don Bunch and I were scheduled to hunt on Dr. Bill Bridgeford's ranch near Maryneal,

*Karen Boggs could catch snakes as well as, or better, than most men.*

and we invited Karen and her husband, David, to join us since our group wasn't very large that day.

We arrived at the Bridgeford Ranch and visited briefly with Dr. Bill and his wife. Dr. Bill gave us directions to several dens on his ranch but had no desire to go with us. He had but one request: "Catch all of those damn things!"

The Bridgeford Ranch covers about seven square miles, so we had plenty of room to hunt. The first promising place we spotted was a large flat boulder not fifty yards from the road. Don and I were in Ol' Tuffy, and Karen and David were in their pickup. I stopped to look around and walked up on a five-foot rattlesnake sunning itself. I caught the snake and went back to Ol' Tuffy to get my spraying equipment. We caught several more rattlesnakes, and Karen joined in as if she'd been hunting rattlesnakes all her life. She wasn't afraid at all.

We left that den and had driven about a mile when we noticed a high bluff with cracks about six feet down from the top. Don said, "There won't be any rattlesnakes here because the bluff faces north."

It's true that you don't see many dens facing north, but I wanted to take a closer look. I pulled my way up the bluff using tree branches and brush growing out of the steep hill. When I reached the opening in the rock, it was packed with leaves, cactus and branches—a sure sign that pack rats had been busy building a snug home in advance of cold weather.

Rattlesnakes and rats den up together in the fall to hibernate, so I used my snake hook to pull what debris I could from the opening and shined sunlight on my mirror toward the back. The den was full of rattlesnakes.

I yelled to Don, Karen, and David that I had found a good den and to drive Ol' Tuffy closer so we could get to our equipment. When Don climbed up to the den and looked in, he said, "Well, so much for the rattlesnakes not building dens that face north." We laughed when I told him I guessed this batch of rattlesnakes had lost its compass.

We misted the den with gas vapor and within two hours had caught seventy-five pounds of rattlesnakes. That brought our total for the day to more than a hundred snakes. We stayed on the

Bridgeford Ranch most of the day and headed back into town tired but armed with new snake tales. From then on, we gave Don a tough time about being expert on where rattlesnakes den up.

Karen and David were members of my regular hunting group for years, and Karen accompanied Bill Whitaker and me on a couple of adventures that resulted in *Abilene Reporter-News* feature columns. Karen may be small in stature, but her courage and spirit of adventure are huge.

# Chapter 14
# Rambo and Lisa

Bob and Lisa Barkley of Carrollton attended their first Roundup in 1986. They arrived in Sweetwater on Thursday but weren't scheduled to go hunting with me until Sunday. They were anxious to hunt and after several sad, pleading looks from Lisa, a couple of Jaycee buddies cornered me to ask if I would take the Barkleys with me on Friday.

I asked about the number in their hunting party. The Barkley group consisted of three Spanish men who spoke no English; two citizens of mainland China who spoke only broken English; Bob, Lisa, and Bob's teenage son. That Friday as I looked across the Coliseum at the group, I noticed that Bob and his son were wearing camouflage clothing with tee shirts that proclaimed, "Kill them all and let God sort them out!"

Attached to his belt, Bob's son had a knife that I swear stretched from his waist to his knee.

After one quick look, I told my Jaycee pals that I would have to pass on the group that day. I already had all the people I could safely guide, but I agreed to take them on a hunt on Saturday. When I left the Coliseum, the Barkleys were still looking for a hunting tour to join.

Early Saturday morning when I arrived at the Coliseum to round up my group for the day's hunt, someone pointed me out to Lisa, and she headed toward me on the run. She introduced herself and said that she and her friends had tried to find me all day Friday. "Really?" I said. She said that they had gone on a Jaycee-led hunt but hadn't found any rattlesnakes and were very disappointed after driving all the way from Carrollton, a Dallas suburb.

I asked if they had brought anything to hunt with, and she assured me that her husband, Bob, had his own snake hunting equipment. When I agreed to take them with me, Lisa ran over to her group and gave them the news. Here they came—three Spanish men, two Chinese men, Bob, Lisa, and a teenager. I never found out how the multicultural group got together.

I cautioned everyone that conditions were dangerous at the Catfish Ranch and went through my safety routine, including the parts about watching where you step and not picking anything up without first checking the surrounding brush.

My group of newspaper and television people, along with the Barkleys and their friends, departed the Coliseum about nine o'clock, headed for the ranch. I led the way in Ol' Tuffy, and two or three more pickups completed our caravan. Lisa rode in the cab of one of the pickups, and Bob and his son climbed into the pickup bed. Neither was wearing a shirt although both were sunburned from the day before.

As we drove the rough, dusty trails through the ranch to the den site, I looked back to check the rest of our caravan. Bob and his son were standing in the back of the pickup bed, letting mesquite limbs hit them on their chests as the pickup knifed through the heavy brush. Bob was yelling, "Good man, good man!"

Bill Fink, who was riding with me, turned and looked at me in a puzzled way and asked who the two men were. I said, "It's going to be a very interesting day."

Bob Barkley is an ex-Marine and in great shape, but I don't believe that being whipped with mesquite limbs was part of his Marine basic training. Others in the Barkley group were more passive, thank goodness. When we arrived at the den, Bob and his son jumped from the bed of the pickup, their upper bodies covered with mesquite limb welts, and began unloading their snake hunting "equipment." I walked over to take a closer look.

Bob had taken an old golf club shank and driven a nail into it, bending the nail to a ninety-degree angle. This was what he had brought to hunt with—nothing else. I told him that I didn't want

*'Rambo' and Lisa with Tom Wideman.*

to hurt his feelings, but his rattlesnake hunting equipment was useless and should be left in the pickup. He reluctantly agreed after I offered to let him use one of my snake tongs.

We got several rattlesnakes out of the den, and Bob caught one. One of the young Chinese men was so nervous that he missed a rattlesnake coiled up, sound asleep, at the mouth of the den. After gently teasing him for letting it get away, we sprayed gas vapors into the den and waited for the rattlesnake to come out.

I was beginning to get tired after hunting all week. I sat down above the mouth of the den while Bob and the rest poked about. I quietly called to Lisa and motioned her to walk over to where I was. The rattlesnake the young Chinese man had missed earlier was coming back out, and I could see its head a foot or so from where I was sitting.

I told Lisa to get some snake tongs so she could catch it. She was nervous, so I asked her to wait until I gave the signal before trying to capture the snake. I waited until the rattlesnake was about a foot and a half away from the den and told her to catch the snake in the middle of its body. Lisa followed my instructions and was thrilled with her catch as most first-timers are.

She said, "Tommy, you're so smart. You knew if we waited, it would come out." I laughed and told her one of my favorite snake tales. I think like a rattlesnake, and since they have the IQ of a chicken, that's how smart I am.

By day's end, Bob and his son had put on shirts and were having a great time, as were the rest of their group. We gave Bob the nickname "Rambo" because of the beating he had subjected himself and his son to earlier in the day. Bob's Rambo image mellowed, and he and Lisa became regulars at the Roundup—fun-loving and a little bit crazy.

# Chapter 15
# Snakes and Steaks

Fort Worth businessmen Gary and Joe Pace and I have been good friends since we were kids growing up together in Sweetwater in the 1950s. In 1985, they invited me to join them for a rattlesnake hunt on their Pace Ranch near Lake Sweetwater. They had invited a group of Fort Worth doctors to the ranch for the weekend and were quick to admit that neither they nor their guests knew the first thing about rattlesnake hunting.

Gary and Joe asked me to bring a couple of experienced snake hunters with me. My Arkansas friends, Don Bunch and Bill Fink, were in town and were glad to join the weekend hunt. When we arrived at the Ranch, Gary and Joe introduced us to their guests. Gary had told them that the rattlesnakes on the Pace Ranch were coming from rabbit holes behind the main ranch house. He told me the same thing, but I didn't think the holes he pointed out looked much like rattlesnake habitats. It *was* their ranch, so we misted gas vapor into the holes, but didn't find any rattlesnakes.

Finally, Gary asked where *I* thought the rattlesnakes were making their dens. From the front porch of the main ranch house, I looked across an open field to a hill about a mile away covered with huge boulders and tall grass. I asked Gary if the hill was on their property, and he said it was. I told him we should hunt for rattlesnakes there.

Don, Bill, and I headed toward the hill with Gary, Joe, and their guests close behind. When we got to the bottom of the hill, I suggested that they wait while Don, Bill, and I scouted the area. It was a very warm day, and I was sure we would find some rattlesnakes outside any dens we might locate. As we walked across the thick growth of leaves and brush, we neared a large dead tree that had fallen to the ground. Two rattlesnakes were coiled up, sleeping in the sun. Moving quietly, we spotted three more nearby.

I motioned for Joe and Gary's group to join us. When they got close, I quietly cautioned them to watch their steps because there were five rattlesnakes in the area. When they reached the spot where we were standing, I asked if they could spot the five snakes. None of Gary and Joe's friends had ever been snake hunting and weren't familiar with a rattlesnake's natural camouflage. After looking closely, the ten people in their group could locate only two rattlesnakes.

I picked up the other three snakes with my tongs. The guests were amazed that rattlesnakes could hide so well in the leaves and dead grass, and they started walking with great caution, inspecting the ground carefully before each step. I had ten snake tongs and a few snake hooks with me, and soon the Fort Worth visitors were using them to poke about, but no one found any-more rattlesnakes in that area.

I went off alone to a huge boulder, which sloped down from the top of the hill at a forty-five-degree angle. Instead of taking time to climb down the hill, I decided to slide down the boulder's smooth face, hoping to check a promising location at the rock's base. As I was descending, I looked toward the bottom of the boulder and spotted one of the largest rattlesnakes I had ever seen, coiled up just below me. I was wearing leather gloves, as I always do while snake hunting, and by the time I halted my descent, I had worn holes in the palms of both gloves! Lucky for me, the rattlesnake was asleep.

Off to one side, I spotted the large den from which it had probably come. I called for some snake tongs, and the whole group ran to the area above me, wanting to get in on the capture. Since no one in the group had caught a live rattlesnake before, I figured they had a slim chance of catching this one before it got away. I asked Don to slide the tongs down to me.

Awakened by the noise and movement, the rattlesnake uncoiled and slithered back toward its den, rattling loudly. By then I had the snake tongs and managed to grasp the rattlesnake by the last foot of its six-foot length just as it entered the mouth of the den. All I could do was hold firmly as the rattlesnake struggled to get away. After a contest lasting about ten minutes, I finally pulled the rattlesnake from the den. I tossed it about ten feet away from the mouth of the den, which gave me time to get down from the boulder and catch it.

Each doctor wanted to have his picture made holding the rattlesnake in the tongs and Joe obliged. Putting the large rattlesnake in my snake box, we continued hunting. Don and Bill found another den nearby, so we misted it with gasoline vapor and caught fifteen more snakes, ranging in length from two to three feet. We caught twenty-one snakes in just over two hours.

By then, the Fort Worth group had "enjoyed all the snake hunting they could stand," so we headed back to the main ranch house. Joe wanted a rattlesnake skin, so I killed one of the largest snakes and skinned it.

Gary had planned a steak dinner. I told him that as a special taste treat for their guests, I would cook the rattlesnake along with the steaks. Out in the cook shack behind the house, I cut the rattlesnake into three-inch pieces and placed them on top of the steaks. When Gary saw what I was doing, he was alarmed, afraid that I was getting "snake poison" on the steaks. I assured him that this wasn't true and that he could leave the cooking in my capable hands. When the steaks and rattlesnake were cooked, I took them back to the ranch house where we enjoyed a good meal with salad, beans, onions, and beer. Cooking the rattlesnake with the steaks turned out great, but Gary wouldn't eat a bite of either. The rest of the group wasn't shy, and there wasn't a piece of steak *or* rattlesnake left over.

Joe had the rattlesnake's skin mounted and uses it for wall decoration in the main ranch house.

# Chapter 16
# The Hot Seat

ABC News sent Ned Potter and a film crew to the Sweetwater Roundup in 1982 to film an Evening News feature story. I was asked to lead them on an up close and personal West Texas rattlesnake hunt. Potter is from Chicago and an experienced reporter, but I don't believe he had ever been given an assignment exactly like this one.

Paul Hill was waiting for us when we arrived at his ranch, and off we went. A cameraman laid claim to the passenger seat in Ol' Tuffy. Ned wanted to ride with us so he could ask questions as we drove to the rattlesnake den we planned to work, so he climbed in the back of Ol' Tuffy and made himself comfortable on top of my snake box.

I had been hunting on the ranch the day before and had already caught quite a few rattlesnakes, which were in my snake box, along with my snake hunting equipment.

As we drove down a pasture road, Ned tapped me on the shoulder. "What's that hissing sound?" he asked.

I pointed to the sturdy wooden box he was seated on and answered, "You're sitting on yesterday's catch of rattlesnakes."

The words had no sooner left my mouth than Ned leaped from Ol' Tuffy, raising a cloud of dust as he tumbled on the ground.

I couldn't believe that Ned had jumped from a moving vehicle. I stopped Ol' Tuffy as quickly as I could and ran to where Ned had raised himself to a sitting position. He was covered with dirt, had torn one of his shirt sleeves, and was obviously upset and angry. I didn't notice that his cameraman had begun filming.

Ned glared at me and said, "Why in the hell didn't you tell me you had rattlesnakes in there?"

I replied, "Well, you wanted to ride in the Jeep, you didn't get bit and, besides, you didn't ask me what was in the box. Anyway, if you'd been bitten on the rear, you would've found out very quickly if you have any friends with you."

He thought about that for a moment and slowly began to laugh. The rest of his crew and I began laughing, and the tension of the moment melted away.

We got back in Ol' Tuffy the same way we had started out, with the cameraman sitting in front with me and Ned in the back sitting on the snake box. This time, however, he covered the top of the box with cardboard.

# Chapter 17
# It's Only Flat on One Side

I had the pleasure of leading a rattlesnake hunt on the Hills' Palava Ranch for an NBC Evening News crew in 1986.

James Koepp, television specialist, and Chuck Pharris, cameraman, were dispatched by NBC to film a documentary on the Sweetwater Rattlesnake Roundup. Their teenage sons accompanied them on the assignment.

The NBC crew arrived at the Nolan County Coliseum in a new GMC Suburban piled high with camera equipment and suitcases—not exactly the best vehicle for the rugged terrain of the Palava Ranch.

At the ranch I urged Chuck, who was driving the Suburban, to stay close behind Ol' Tuffy because the den was in the middle of a pasture about a mile and a half off a "road" cutting through the ranch. Paul Hill led the way in his pickup, and Don Bunch rode with me in Ol' Tuffy.

When we arrived at the den and got out of our vehicles, I cautioned Chuck's son, who was wearing tennis shoes, to watch where he stepped since it's extremely dangerous to hunt rattlesnakes without wearing some type of sturdy boots for protection.

At the mouth of the den, we saw five or six rattlesnakes coiled up in front. We waited while the camera crew quickly unloaded its equipment, and Don and I moved in quietly to pick up two or three rattlesnakes. As we were doing this, the other rattlers slithered rapidly back into the den. We sprayed a fine mist of gasoline vapor into the back of the den and waited. I figured that rattlesnakes coming out first would break into the open quickly, and I wasn't disappointed.

Chuck's son was standing next to me when two rattlesnakes emerged from the den simultaneously. He had snake tongs, but he moved quickly in the opposite direction from where his dad was filming and called over his shoulder, "Even back here is too close for me!"

I caught both rattlesnakes. Others began slithering out from different openings in the den. Don, Paul and I picked them up and put them in my wooden snake box while the cameras rolled.

We worked two or three more dens and, by day's end, Chuck's son was catching rattlesnakes like a pro.

The NBC crew filmed almost six hours of material to be edited for an Evening News broadcast the next day. I asked James Koepp who would narrate the final segment. He said Charles Murphy, who was to arrive in Sweetwater the next day, would do the honors.

Don and I piled our snake hunting equipment into Ol' Tuffy while James, Chuck, and their sons loaded camera equipment into their Suburban. Paul had already headed back to the Coliseum in his pickup. I cautioned Chuck to follow closely behind Ol' Tuffy in my tire tracks. Cedar trees had been pushed down in the area, and there were some sharp stumps that could cut through a tire like a knife.

In Ol' Tuffy, Don and I got to talking, reviewing the day's events as we drove. When we arrived at the main gate of the ranch, we looked back. The Suburban was nowhere in sight. We waited at the gate for fifteen minutes and when they still hadn't arrived, we drove back toward the last den we had worked.

A mile or so back into the pasture, we saw the Suburban in an area completely off the beaten trail. Its left rear tire had a five-inch cut in it where Chuck had run over a cedar stump. They were hot and sweaty from having unloaded their cargo of camera equipment and luggage to get to the spare tire. Their tire jack wouldn't support the weight of the car because of the soft dirt, and Chuck's sense of humor had completely disappeared. They were in the middle of nowhere with a flat tire and a non-functioning tire jack.

When I asked Chuck why he had driven off the trail against my warnings, he replied, "Hell, I don't know. I guess I don't hear so good!"

I let him fume for a minute, and then told him I had a hydraulic jack in the back of Ol' Tuffy that I "rented by the minute." He laughed finally, and soon the Suburban was up, the ruined tire removed, and the spare in place.

Don and I, being gracious West Texas hosts, sat in the shade of a nearby mesquite tree and watched the flurry of activity, dispensing *advice* as necessary. Chuck and James knew they were in for some kidding, and we didn't disappoint them. When we arrived in Sweetwater, the Coliseum had closed for the day, so we made plans to meet them there the next day—Saturday.

Don and I took another press group out Saturday morning, but were back at the Coliseum by early afternoon. We located Chuck and James, who introduced us to Charles Murphy. Mr. Murphy said Chuck had told him how *generous* Don and I were to offer verbal assistance while they changed the flat tire and added that he would have joined us under the shade tree had he been with us.

Later that afternoon, Mr. Murphy filmed his segment from the floor of the Coliseum. Dressed in an African safari jacket and blue jeans with a red bandana tied about his neck, he told the television audience, "Well, I'm here in Sweetwater, Texas, at the World's

Largest Rattlesnake Roundup. We've been shaking, rattling, and rolling over the West Texas plains today, catching live rattlers. I've just returned from the field where this afternoon our hunting party caught over a hundred pounds!"

I was amused by Mr. Murphy's comment about hunting snakes "in the field." He hadn't been within fifty miles of a rattlesnake den, but he certainly knew how to dress the part! When the segment aired on NBC the next day during the Sunday Evening News, Sam Donaldson did the lead-in. From six hours of filming, the Evening News segment covering the Roundup lasted about two minutes.

Even though NBC used only a small portion of the rattlesnake hunt footage for their Evening News segment, it didn't wind up on the cutting room floor. Portions of the film were broken into short features and sold to various airlines for viewing during intercontinental flights.

Gene and Margaret Day, friends from Fort Stockton, were en route to Hawaii on American Airlines in 1986, when a thirty-minute sports feature was shown before the in-flight movie. To their surprise, the Sweetwater Rattlesnake Roundup was highlighted. Gene and Margaret went up and down the airplane aisles after the movie, telling anyone who would listen that they knew the crazy guy in the rattlesnake feature.

# Chapter 18
# The Two-Headed Cow

When a national cable network assigned a reporter to cover the Sweetwater Roundup in March 1987, they contacted my hunting buddy, Tom Henderson, who manages the Blue Goose Ranch in Nolan County. Tom made the arrangements for the visit and invited fellow Jaycees Richard Eschugen and me to accompany them.

The day the crew arrived, I was running behind schedule and missed them at the Coliseum by several minutes. Tom had told me where they would be filming, and I caught up with them at the Blue Goose Ranch near a den we had named Red Bluff Rattler

Hotel. Red Bluff is a rattlesnake haven with many dens—some too deep to get even the longest piece of copper tubing to the back of the den.

When I arrived at Red Bluff, the crew was already filming. From the sidelines, I surmised that this was the reporter's first encounter with live snakes—rattle or otherwise. He wanted the filming to feature his one-on-one encounter with a rattlesnake. He was to be the star of the show, so we prepared to act out our parts.

Tom gave him some background on rattlesnakes and their habitats and a brief overview of what happens when gasoline vapor is misted into the den. From this information and our laid-back attitudes, the reporter apparently assumed that with a snake catcher in one hand and a mirror in the other, he would be in complete control of the situation. The rest of us watched, mulling over in our minds what to expect from him should something unexpected happen.

Tom sprayed the den with gasoline vapor, and then he and the reporter made small talk about the snake den, what spraying procedures Tom had used, and rattlesnakes in general.

I noticed two rattlesnakes coming out of a hole that hadn't been misted with gasoline, and since neither Tom nor the reporter seemed to be aware of them, I walked over to pick them up with my snake tongs. The reporter yelled at me to "get out of the picture!"

Meanwhile, Tom noticed a four-foot rattlesnake easing its way out of the den. He called the reporter to come over to the mouth of the den to capture the rattlesnake. I could tell that his "moment of truth" had arrived by the way he tiptoed his way toward the mouth of the den, dressed in his snake hunting regalia, which included low-quarter deck shoes.

With Tom standing by him, the reporter appeared to be a fearless snake hunter. At that moment, the rattlesnake bolted out of the den toward him and he jumped back and ran ten or fifteen feet, visibly shaken. Tom, who hadn't moved an inch, began laughing.

Filming stopped while the reporter composed himself. With Tom's encouragement, he walked slowly back to the den although the rattlesnake was now gone, heading up the hill, disappearing into the brush.

Filming began again. As the reporter approached the rattlesnake, Tom coached him to catch it at the middle of its body. But he squeezed so tightly that the rattlesnake started thrashing around. He dropped the snake and headed off again in the opposite direction, even though he knew that if he was going to be the star of the show, it was now or never.

Tom had already caught the snake, but he turned it loose so the reporter could demonstrate for the camera how to capture it. With cameras rolling, on his third attempt he picked up the rattlesnake and placed it in a snake box for safekeeping. It was obvious that he wasn't enjoying this adventure and that one rattlesnake was more than sufficient for his purposes.

Several days earlier Tom had found a complete skeleton of a cow. Being a mischievous West Texan, he moved it to the vicinity of the Red Bluff den the morning of the crew's visit. An unusual feature of this cow skeleton was: it had two heads!

The reporter noticed the skeleton and asked for details. Tom told him with a straight face that the cow had been bitten by rattlesnakes and had died where it fell. The reporter's expression was equally somber as he listened.

Richard and I didn't smile; we just listened quietly to Tom's story. Tom went on and on about how sorry he was to lose the "special cow" born with two heads. He said he had hoped to get a herd of them started because they gained weight twice as fast as other cows, being able to eat with both heads. The reporter either wasn't listening closely or was still in shock from his rattlesnake catching episode. You could tell he was accepting Tom's story as truth.

Suddenly, he noticed someone in our group grinning from ear to ear. Realizing that he was being kidded big time, he tried to pretend that he had known all along it was a joke. He quickly ended Tom's interview and declared the day's filming officially over.

When the Rattlesnake Roundup segment aired, some of the most interesting parts had been deleted. The footage of the reporter catching the rattlesnake was intact, but the parts where he fell down and ran off were missing, and Tom's two-headed cow interview also had been deleted.

# Chapter 19
# National Geographic Explorer

In January 1991, I received a telephone call which would forever change my life.

Kevin Bachar, post production supervisor for the National Geographic Explorer program headquartered in New York City, contacted the Sweetwater Chamber of Commerce seeking the name of someone who had been actively involved in the Jaycee Rattlesnake Roundup for a long time. The chamber gave him my name. A few days later, he telephoned me in Abilene. During our conversation, I learned that the National Geographic staff was considering an Explorer segment to be filmed at the 1991 Rattlesnake Roundup.

Kevin asked, "Do you believe the activities of the Sweetwater Rattlesnake Roundup could hold the attention of the Discovery Channel's world-wide audience?"

Did I ever! We spoke for more than an hour, and I like to think that it was my sincere enthusiasm for the project that finalized his decision.

Kevin and his crew arrived on March 4—Peter Schnall, producer/cameraman; his sister, Lisa Schnall, sound technician; and Rick Joya, lighting engineer.

When they were settled in their motel, I told the group how pleased and excited I was that they had come to film the Roundup. Kevin asked if we could get an early start the next morning. I asked, "How early?" and he answered, "Sunrise. I want to get a shot of a West Texas sunrise and a West Texas sunset."

I told him, "Sunrises and sunsets look alike out here. I'll pick you up about 9 a.m."

The next morning at nine o'clock as promised, Bill Whitaker and I arrived to find the National Geographic crew waiting by their van. I asked the group if they had rested well and, before any one else could answer, Lisa launched into her tale about the black snake.

***Tom Wideman was interviewed for National Geographic Explorer documentary. (Photo by Bill Whitaker)***

Shortly after I left them the evening before, Kevin, Rick, Peter and Lisa went to their respective rooms. All were tired from their New York to Texas flight and knew the remainder of the week would be very busy.

Lisa said that it was her custom to take her shower at night before going to bed. When she pulled the shower curtain back to turn on the water, she spotted a four-foot snake stretched out like a black exclamation point on the white porcelain. She screamed so loud that other motel guests within earshot, including her brother, Peter, and Kevin and Rick, rushed out of their rooms to see what had happened.

When she saw the big grin on Peter's face, she realized who had put the *rubber* snake in her tub. She told her brother to laugh while he could because she would get him back. It was almost 9:30, and Kevin reminded us that we had lots of filming to do that day. Soon we headed for the Blue Goose Ranch in search of a rattlesnake den.

When we arrived at the gate of the Blue Goose, my snake hunting buddies Tom Henderson, Richard Eschugen, and David Boggs were waiting. Everyone began unloading gear, preparing for the hunt. I could tell that Lisa was still very apprehensive as she pulled snake-proof leggings up over her blue jeans. She walked to where I was and asked, "Are there any other poisonous snakes out here besides Western Diamondback rattlers?"

I couldn't resist. "The only other dangerous snake we have in West Texas is the Western Diamondback Hoop Snake," I said. She asked for a description of the Hoop Snake.

"Well," I said, "if you look up one of these trails and see something rolling toward you that resembles a bicycle tire, and it's kicking up lots of dust, you'd better run because it's a Hoop Snake for sure!"

"What happens if you can't outrun it?" Lisa asked.

"It rolls right up to you, straightens out, and bites you on your behind!"

Lisa's eyes got as big as saucers, and she turned to her brother and said, "Did you hear what Tommy just told me, Peter?"

He replied, "Lisa, remember, they *never* tell a lie in Texas."

Realizing that she had been tricked again, Lisa turned back to me and said with a laugh, "Darn you! Now I'll have to get even with you, too!"

Lisa was beginning to realize that her brother and I were giving her more trouble than any rattlesnake she might encounter that day, so she put her dread of the day's hunt behind her.

The Blue Goose Ranch covers seven sections—seven square miles. We headed for the Red Bluff rattlesnake den, which has lots of crevices and holes, some over twenty feet deep. There are *always* rattlesnakes at Red Bluff, so it was a natural place to begin. When we arrived at the den, seven or eight rattlesnakes were warming themselves in the early morning sun. I asked the film crew if they could see the snakes, but neither Kevin, Peter, Lisa, nor Rick could spot one. From a distance, I pointed out the rattlesnakes' locations, one by one.

Before we started working the den itself, Kevin said that he wanted to do an on-camera interview with Tom Henderson and one with me. After the interviews were filmed, we turned our attention to catching the rattlesnakes still coiled in front of the den.

Kevin positioned his Explorer crew and then told Tom and me what he wanted to film. With Peter's camera rolling and Lisa capturing every sound on tape, we quickly caught the rattlesnakes we had seen when we arrived at the den.

We sprayed the back of the den with a light vapor mist of gasoline, using copper tubing attached to a spray can. Although this procedure is a controversial part of snake hunting, the amount of gas vapor we use is so small it does no lasting damage to the den. We return year after year to the same dens and continue to find them filled with rattlesnakes. It's not unusual for snake dens to branch out in many directions, so rattlesnakes began to slither out of several holes around the den.

We worked the Red Bluff den for almost an hour and caught about twenty rattlesnakes, mostly three-footers. However, Peter had a special camera shot in mind and wanted the largest rattler we could find. We loaded our gear, left the Red Bluff, and headed to a rugged area where we had caught some very large snakes in past years.

Unloading the snake hunting gear again, we walked toward a gully filled with large rocks, brush, and dead tree limbs. In order to film a panoramic view of the action, Kevin and Peter set up their cameras across from the rock gully where Tom, Richard, and I were looking for a promising den.

Tom was walking about ten or fifteen feet ahead of me. As I followed almost in his footsteps, I looked down and saw a big five-foot rattlesnake coiled in the grass. Tom had walked within two feet of this big fellow and hadn't seen him. The rattlesnake was dozing. I froze in my tracks and quietly called to Kevin that I believed I had found the large rattlesnake that Peter wanted for his special film segment.

The Explorer crew was trying to reposition its cameras when the big snake woke up, rattling loudly, and began to crawl into the dense underbrush. I was sure it was going to get away before Peter could get his cameras set up, but Peter called over to me that he could film using his telephoto lens and for me to proceed with catching the snake.

My immediate problem was that the rattlesnake was well over five feet long and I was carrying snake tongs only two feet long. There wasn't time to lose, so I reached for the snake with my tongs, grasping him about the middle of his long body. The rattlesnake immediately started striking at my legs, twice missing me only by inches.

Richard said, "Wow, that's a man's snake!" Then, realizing that I could use some assistance, Tom Henderson said, "S—- fire, Tommy, those tongs are too short!"

Knowing that I wasn't about to let that big rattlesnake get away, no matter what, Henderson walked quickly toward me carrying four-foot tongs, and I passed the snake, which was still striking wildly, to him.

It wasn't until I viewed the final cut of the Explorer film that I realized how close I'd come to being bitten twice on my right thigh. At the time, however, we were too busy to get upset.

Kevin and Peter had captured the entire exciting episode on film.

# Chapter 20
# Snake-Proof Boots

The Gokey snake-proof boots I wear during the Roundup, and whenever I hunt, have opened doors to interesting friendships and travel opportunities.

Gokey Boot Company, which has a large manufacturing plant in Tipton, Missouri, is owned by the Orvis Company, an elite sporting goods manufacturer and retailer headquartered on the East Coast.

Gokey boots have long been revered by sportsmen, including Dwight Eisenhower, Ernest Hemingway, and other well-known outdoor enthusiasts. The boots, which are made individually by

*Tom wears his Gokey Boot Sauvage and holds the all leather Gokey boot he has worn for many years while hunting or handling snakes.*

hand, are incredibly strong, yet comfortable. They're expensive, but when you consider that a snake bite can cost you from $10,000 to $50,000 in medical expenses and can result in the loss of a finger, hand, toe, or leg, the boots are cheap by comparison.

My boots have been stuck no less than three hundred times, and I swear by their protective qualities.

Because of my association with the Rattlesnake Roundup, and because I've worn the same pair of Gokey all-leather boots since the early 1980s, I wrote a letter to the Orvis Company's CEO, Perk Perkins, in the fall of 1992, chronicling my experiences while wearing their snake-proof boots. In my letter, I mentioned that my Gokey boots had been featured prominently in the 1991 National Geographic Explorer coverage of the Sweetwater Rattlesnake Roundup.

In a couple of weeks I received a response from Mr. Perkins, and in January 1993 he wrote that he was sending Jerry Doss, manager of Gokey's Missouri plant, and his son, Stacy Doss, to that year's Roundup.

These two men, who design Gokey's line of luggage, shoes and boots, including the snake-proof boot or 'Boote Sauvage' as it's listed in the Orvis catalog, had recently designed and handcrafted a new snake-proof boot, constructed with a green, space-age ballistic material top reaching to the knee, combined with traditional leather bottoms. With the new design, the boots are a couple of pounds lighter than the original all leather model—an important consideration to any hunter who has ever spent a long day hiking through rough country.

Mr. Perkins asked if I, as a loyal, twenty-year fan of Gokey's all leather boots, would be interested in field testing Jerry's and Stacy's new design. The company had been testing the new boots in its Tipton plant laboratory, but we agreed that they had not been tested scientifically until they had been worn during an actual rattlesnake hunt.

A week later, I received a pair of the prototype boots in the mail. The Gokey Boot Company custom fits every pair of shoes and boots it sells and keeps a file for each of its customers, detailing this custom fit information. When the prototype boots arrived, they had been fitted to my specifications. I put them on, and they were as comfortable as my 1982 vintage all leather boots.

In March, Jerry and Stacy Doss arrived in Sweetwater and spent Friday and Saturday rattlesnake hunting with me. That year's Roundup brought in ten thousand pounds of rattlesnakes in spite of the fact that it rained all day Saturday. Because of the inclement weather, Jerry, Stacy, and I cut short our snake hunt Saturday afternoon and returned to the Coliseum to find it filled to overflowing with interested spectators, agitated rattlesnakes, and busy Jaycees.

Jerry and Stacy said they had never seen anything like it. They couldn't believe the crowds in and around the Coliseum or the excitement generated by close proximity to thousands of live rat-

*Decked out: Tom Wideman sports Orvis-Gokey gear and a bolo tie given him by a friend.*

tlesnakes. We made our way through the crowd to the weigh-in area located near an eight-foot by eight-foot holding pit, which that afternoon held three to four thousand writhing rattlesnakes.

I invited Jerry to step into the pit with me. He and Stacy had worn the traditional all leather Gokey boots since their arrival in Sweetwater, and I thought it would be an excellent opportunity for Jerry to see how well they performed under pressure.

Jerry quickly made it clear that he wanted me to get into the pit before he did. I said, "You're not even wearing the prototype boots I'm field testing for you. Why are you concerned about getting into the pit wearing your all leather boots?"

Jerry said, "You jump over in there and let me see what happens," and so I did.

The prototype boots deflected the striking rattlesnakes every bit as well as my all leather boots. Even with this demonstration of the prototype boot's snakeworthiness, however, neither Jerry nor Stacy were willing to join me in the rattlesnake pit that rainy afternoon. Several years later on another rainy afternoon in Austin, Jerry did voluntarily get into a rattlesnake enclosure with me, but that's another story.

Several months later I received another call from Perk Perkins. This time he invited me to join him for the grand opening of a new Orvis store in an upscale shopping center in Highland Park, a suburb of Dallas.

Would I be interested in demonstrating the durability of the Gokey Boote Sauvage in an enclosure with live rattlesnakes? Even though I had no idea what I might be getting into, I was excited about the prospect of being associated with the Orvis Company.

I told him I would need a large enclosure with some sort of walls to keep interested spectators at least a couple of feet back from me and the rattlesnakes. He said Jerry and Stacy would devise a safe and sturdy enclosure.

The day before the store opening, June and I loaded our Ford Bronco with my Gokey boots, several snake tongs and hooks, a snakebite kit and, of course, a large wooden box filled with live Western Diamondback rattlesnakes, and headed for Dallas.

When we arrived at our hotel and went inside to register, I decided against including the box of live rattlesnakes with our luggage—after all, we weren't in West Texas anymore. I left the rattlesnakes in the back of the Bronco locked inside a wooden box clearly marked "Danger! Live Snakes! Do not touch."

Amazingly, the bellman who assisted us with our luggage didn't comment on the snake box. Maybe he thought it was a joke or didn't really want to know.

We drove to the Highland Park store to meet with Jerry Doss and Mr. Perkins. The store was outfitted with sporting equipment of every description, fine clothing, luggage and travel cases, fly fishing rods and accessories—everything a serious sportsman could desire. Demonstration tables were arranged throughout the store, including a table Jerry had set up where he planned to custom make a pair of Gokey boots beginning with nothing more than a large piece of uncut leather.

Outside, I found a ten-foot-square tent had been set up for me. On the floor of the tent were four two-by-six foot pieces of plastic sheeting, four plastic poles two feet tall, and several rolls of duct tape, which meant that my demonstrations would be done within a six-foot square enclosure. I knew that these weren't ideal circumstances, but I'm comfortable working with rattlesnakes and would be handling only one snake at a time.

With June's help, I set to work building my enclosure. June asked, "Are you sure this setup is going to work?" I smiled and assured her that people from Vermont weren't familiar with rattlesnakes and that's why they had invited me.

Bright and early on opening day, Friday, people filled the new Orvis store and began to overflow into the parking lot and crowd around my demonstration area. The grand opening brochure stated that I would do demonstrations at 10 a.m., 2 p.m., and 4 p.m. I did my first demonstration at 10 a.m. but didn't get out of the enclosure until after 6 p.m. It was non-stop show time. I let the rattlesnakes strike my Gokey boots while answering hundreds of questions about the boots, rattlesnakes, and how many times I had been bitten by a rattlesnake.

About 2 p.m. a black stretch limo pulled up and a tall, well-dressed gentleman stepped out, accompanied by two people who were obviously body guards. Even though I had put bright yellow Do Not Cross tape around the demonstration area, he lifted the tape, greeted me with a warm handshake, and introduced himself as Steve Bartlett, mayor of Dallas.

I told him that I had been mayor of Sweetwater and that if he served as Dallas's mayor as long as I had served as Sweetwater's, he would wind up chasing rattlesnakes, too.

"Heck," he replied, "I've been chasing rattlesnakes for a long time here in Dallas, but they all have two legs!" The crowd laughed and applauded.

Someone must have advised Mr. Perkins that the mayor had arrived because he took him inside for a photo opportunity. In a little while, Mayor Bartlett returned to shake my hand and tell me how much he had enjoyed our meeting. We traded cards, and he got back into the limo, smiled, and waved as he drove away.

With the added publicity of the mayor's visit, I knew Saturday would draw another big crowd. When I arrived at the store before 8 a.m., people had already surrounded my tent two or three deep. It was a warm day, and the rattlesnakes were very active, so we had a steady crowd of onlookers.

At 4 p.m. Perk and his brother, David Perkins, came out to my area, and I introduced them to the crowd. I invited Perk to join me in the pit, but he politely declined. Several spectators called out, "I'll do it!" David was wearing a pair of all leather Gokey boots, and he accepted my invitation.

As he stepped into the pit, a four-foot snake hit the toe of his left boot. The crowd went wild and applauded him when he exited the pit a few minutes later, smiling from ear to ear. I did nonstop demonstrations until 6 p.m. when I locked the exhausted rattlesnakes in my snake boxes, and June and I went inside the store. About 8 p.m., the day's sales were totaled, and David announced that they exceeded the one day total of any Orvis store in the country.

Perk Perkins called again in August 1998 and invited me to demonstrate the Gokey snake-proof boots at the mid-September grand opening of an Orvis store in Austin.

The announcement brochure he sent read: "Are Gokey snake-boots snake-proof? Tom Wideman, professional snake handler, and Jerry Doss, General Manager of Gokey, will be on hand with live rattlesnakes to put the boots to the test! Demonstrations at 12, 1:30 and 3."

Remembering the fun of the Orvis store grand opening in Highland Park, and with immense pleasure at having been designated a "professional snake handler" by such a respected sports equipment franchise, I gathered my gear and several good-sized Western Diamondback rattlesnakes, and June and I headed for Austin. We arrived to a steady rain falling across the city.

At the mall, we spotted a small tent on the narrow parking area near the Orvis store entrance. Inside the tent was a Plexiglas square, six feet by six feet, with sides four feet high. The tent had been set up on an incline, and a steady stream of water ran across the asphalt surface of the parking area that served as the floor of the tent.

Even though my first demonstration wasn't scheduled until noon, by 10:30 a large crowd of onlookers, including customers and Orvis personnel, had gathered around the tent. They crowded together in the narrow space between the Plexiglas square and the tent's awnings in a futile attempt to escape the rain.

In compliance with a City of Austin ordinance concerning crowds of people and live rattlesnakes in close proximity, the Orvis Company hired two paramedics with an ambulance to stand by during the grand opening. As I finished unloading the snake boxes, the paramedics joined the group of onlookers crowding around the tent.

The senior paramedic informed me that neither he nor his assistant had ever been around rattlesnakes and had never treated anyone suffering from a rattlesnake bite. He asked me, "If you are bitten, what do you want us to do?" My response was swift and emphatic. "If I get bitten, please, don't touch me!"

My demonstrations began shortly after 10 a.m. and continued without a break throughout the day. About 2 p.m., Jerry Doss volunteered to help me in the rattlesnake demonstration area. I had several large rattlesnakes that I was alternating frequently because they were getting soaked as they coiled on the pavement floor of the tent. Rattlesnakes aren't fond of water anyway, and these critters were already aggravated about being poked and prodded to strike for the entertainment of the crowd. Jerry stepped into the Plexiglas square, picked up some snake tongs, and asked how he could help.

I asked him to hold the rattlesnake I had been using down on the pavement while I lifted another relatively dry rattlesnake from one of my snake boxes. I cautioned him not to lift the snake, just to hold it down out of my way.

Jerry pinned the rattlesnake by its middle rather than near its head, and as I reached to lift another rattlesnake from the box, the rattlesnake Jerry was holding grazed the top of my hand with the bottom of its "chin." There's no doubt in my mind that had the rattlesnake been dry instead of damp and sluggish because of the rain, it would have struck my hand.

The crowd gasped, as did June, who had walked up to see how the demonstrations were going. I looked at Jerry, and he returned my glance with a slightly shocked look. I took my snake tongs and lifted the rattlesnake that had grazed my hand into one of my snake boxes, locking the box latch after it slithered inside. Spectators began to ask if I had been bitten. I answered that I hadn't, but that they would *not* see that part of the demonstration again. Jerry and I concluded that the Plexiglas space was too small for two snake handlers.

The crowd around the tent increased as word of my near miss circulated, and it was standing room only the rest of the afternoon. When the Orvis staff closed the store at 6, customers inside joined the crowd outside around my demonstration pit. At 7, I loaded my gear and the rattlesnake boxes into my truck, so the crowd would go home.

During the Dallas and Austin store openings, I was asked numerous times, "What are you going to do with the rattlesnakes now that the demonstrations are over?" People ask me to kill the rattlesnakes, cut the rattles off, and give them as souvenirs, which I don't do.

One guy asked if he could feel one of the snake's rattles, so I pinned a snake, tightly grasped its head in one hand and held its tail and rattles toward the guy. He grabbed the rattles, snapped them off, and disappeared into the crowd. He didn't harm the rattlesnake, but it was a cowardly thing to do.

Chapter 21

# Don't Bite the Hand that Feeds You

One day in July 2004, an Abilene friend called and asked if I wanted to adopt a "pet" rattlesnake.

Her family had provided a habitat for the rattlesnake since it was very small, keeping it in a glass aquarium. Their son had constructed a small wooden house, artfully covered with small multicolor rocks, and Evil, as they named him, slept coiled up in the house when he wasn't moving about in the aquarium.

During the nine years in their care, Evil had grown from about eight inches long to an impressive three feet and had increased in weight tenfold.

The first time I saw Evil, he appeared to be near death—his skin flaky and sallow looking, lethargic and unresponsive, making no attempt to strike at movement outside the aquarium. Having a soft spot in my heart—and my head—for rattlesnakes, I agreed, with June's consent, to adopt Evil.

Because of its distinctive dark red/beige color, we renamed the rattlesnake Red Rider. Rattlesnakes aren't "evil;" they're just rattlesnakes, doing what they were created to do.

We cleaned his aquarium habitat and covered what had been a dirt and wood shaving floor with green Astroturf. We filled his water bowl with fresh water and began offering Red Rider large fuzzies—medium-size white mice, flash-frozen to retain their flavor—purchased from a pet store.

The first two days, he downed six fuzzies, but spent most of the day and evening coiled up inside his rock house. When I scooted the rock house out to the center of the aquarium with Red Rider inside and lifted up the house, he retained his coiled shape, exactly the same dimensions as the house. I added one-inch wooden boards to the bottom of the rock house, raising it to accommodate his larger size. Red Rider was never in danger of escaping because we kept heavy weights on the aquarium lid and dropped fuzzies through a small trapdoor in the heavy wire mesh lid.

About a month later, Red Rider began to regain his color and strength but continued to display no interest in striking, either at the fuzzies—a rattlesnake's natural tendency when searching for food in the wild—or at movement outside the aquarium.

It appeared the rattlesnake had sustained damage to its jaw because its venom sacs were small and misshapen. I could scoot the rock house to the center of the aquarium, lift it up with Red Rider coiled inside, and rub the top of his head, while he lay motionless.

During the winter of Red Rider's stay, we put him and his aquarium habitat inside our pool house, which we keep warm to protect the pool pump mechanism. One day when it was very cold outside, June went into the pool house to get something, forgetting that Red Rider was there. Red Rider happened to be outside the rock house, and they spotted each other at the same moment. June jumped back, and Red Rider instantly retreated to the back of the aquarium into a striking position, rattling loudly. It was the first and only time we ever saw him take this defensive position or rattle.

When it began to warm up, we moved Red Rider and the aquarium to a waist-high sturdy wooden surface on our backyard patio which received morning sun and afternoon shade. One warm afternoon in May we found skin shed on the floor of the aquarium along with two small white fangs. Red Rider's color was now bright and beautiful, and he exhibited good muscle tone down the length of his body, even in the tail and rattle area which

***Red Rider in his aquarium home.***

had appeared almost dead when we first saw him.

Red Rider continued to display the docile temperament I had come to expect and would eat a fuzzie when one was offered—not striking at it, just approaching slowly and then swallowing it lazily. Although June expressed dismay at having Red Rider as a patio guest, I continued to believe that because he had been raised in captivity and had never had to search for food, he would never display the normal tendency to strike defensively.

June kept insisting that "a snake is a snake is a snake." Through the years she has reminded me on several occasions that while

***Tom liked to rub Red Rider's head until one day when the rattlesnake bit him.***

the rattlesnake I'm dealing with at any given time may be number one thousand in my experience, I may possibly be the rattlesnake's first interaction with a human.

In July, the anniversary of Red Rider's arrival at our house, one Saturday morning, as had become my routine, I removed the lid from Red Rider's habitat and, using my left hand, scooted the rock house with him coiled inside, to the middle of the aquarium. As I lifted up the house, I noticed that Red Rider had been coiled inside the house with his head facing toward my hand.

Before I could react, Red Rider struck out and nicked me with one fang on the bony part of my left thumb before retreating to the far side of the aquarium.

At first I thought he had just grazed my skin, but I noticed a small spot of bright red blood beginning to form on my thumb. Muttering to myself, "Well, *this* isn't good," I carefully put the aquarium lid securely back in place and went inside to get the shocking device I had seen used so effectively.

For several hours following the bite, using a clockwise motion, I shocked the area where his fang had penetrated my skin, while counting my blessings that Red Rider had nicked my thumb bone instead of the fleshy part of my hand between the thumb and forefinger. In that area, a nearby vein would quickly have carried up my arm the small amount of venom he had been able to inject.

I will testify to the shocking device's effectiveness. I experienced no damage to my left thumb or upper arm and now have only a small raised bump on my left thumb bone to document my experience. After forty years of snake hunting, I had finally been bitten—by a pet!

Not wanting to alarm June, I made no mention of the incident. However, because we had planned to go to garage sales later that morning and I had suddenly lost all interest in doing so, June was curious as to what might have happened. Finally, later in the day, I told her what had happened. To my eternal gratitude, she didn't say, "I told you so."

I transferred Red Rider from his aquarium habitat to a small, sturdy wooden box and escorted him out of town to an area where a less-traveled road reached its dead end. I turned him loose and watched until he slithered away and disappeared into an area of heavy mesquite tree brush and knee-high weeds.

On warm summer evenings, June and I sit on our patio, wondering how Red Rider is spending his days and if he ever thinks about the little rock-covered house he loved so well.

# Popular Western titles from State House Press

ISBN 1-880510-84-7
$24.95 cloth

ISBN 0-938349-40-6
$14.95 paper

ISBN 1-880510-85-5
$14.95 cloth

ISBN 1-880510-19-7
$19.95 paper

ISBN 0-938349-12-0
$18.95 paper

ISBN 1-880510-58-8
$24.95 cloth

ISBN 1-880510-02-2
$18.95 paper

These books available at booksellers or through
Texas A&M University Press Consortium at
1-800-826-8911 or on-line at www.tamu.edu/upress

# Other titles in the Texas Heritage Series

ISBN 1-880510-81-2
$17.95 cloth

ISBN 1-893114-43-0
$18.00 cloth

ISBN 1-893114-38-4
$14.95 cloth

ISBN 1-880510-82-0
$18.00 paper

ISBN 1-880510-95-2
$19.95 cloth

ISBN 1-880510-91-X
$18.95 cloth

ISBN 1-880510-86-3
$17.95 cloth

These books available at booksellers or through
Texas A&M University Press Consortium at
1-800-826-8911 or on-line at www.tamu.edu/upress